STUDIES IN ECONOMIC AND SOCIAl

CH00690127

This series, specially commissioned by tl
provides a guide to the current interpre
economic and social history in which advanc... ... .
in which there has been significant debate.

Originally entitled 'Studies in Economic History', in 1974 the series had its
scope extended to include topics in social history, and the new series titles,
'Studies in Economic and Social History', signalises this development.

The series gives readers access to the best work done, helps them to draw
their own conclusions in major fields of study, and by means of the critical
bibliography in each book guides them in the selection of further reading.
The aim is to provide a springboard to further work rather than a set of
pre-packaged conclusions or short-cuts.

ECONOMIC HISTORY SOCIETY

The Economic History Society, which numbers around 3000 members,
publishes the *Economic History Review* four times a year (free to members)
and holds an annual conference. Enquiries about membership should be
addressed to the Assistant Secretary, Economic History Society, PO Box
70, Kingswood, Bristol BS15 5FB. Full-time students may join at special
rates.

STUDIES IN ECONOMIC AND SOCIAL HISTORY

Edited for the Economic History Society by L.A. Clarkson

PUBLISHED TITLES INCLUDE

OTHER TITLES ARE IN PREPARATION

Agricultural Regions and Agrarian History in England, 1500—1750

Prepared for
the Economic History Society by

JOAN THIRSK
Sometime Reader in Economic History
in the University of Oxford

MACMILLAN

First published 1987 by
THE MACMILLAN PRESS LTD
Houndmills, Basingstoke, Hampshire RG21 2XS
and London
Companies and representatives
throughout the world

ISBN 0–333–19158–7

A catalogue record for this book is available
from the British Library.

Printed in Hong Kong

Reprinted 1993

Series Standing Order

If you would like to receive future titles in this series as they are published, you can
make use of our standing order facility. To place a standing order please contact your
bookseller or, in case of difficulty,write to us at the address below with your name
and address and the name of the series. Please state with which title you wish to
begin your standing order. (If you live outside the United Kingdom we may not have
the rights for your area, in which case we will forward your order to the publisher
concerned.)

Customer Services Department, Macmillan Distribution Ltd
Houndmills, Basingstoke, Hampshire RG21 2XS, England

Contents

List of Maps

Acknowledgements

I wish to thank Professor Eric Kerridge and Allen and Unwin Ltd for allowing me to reproduce his map of regions from *The Agricultural Revolution*, and Professors L. A. Clarkson and Gordon Mingay and Dr Victor Morgan for reading my manuscript and making many helpful suggestions for its improvement. The author and publishers also wish to thank Cambridge University Press for permission to reproduce two maps from the *Agrarian History of England and Wales*, ed. Joan Thirsk, Vols IV and V.

Notes on References

References in the text within square brackets relate to the numbered items in the Bibliography, followed, where necessary, by the page numbers in italics, for example [1,45]. The other references, in round brackets and numbered consecutively, relate to items in the Notes section.

Editor's Preface

When this series was established in 1968 the first editor, the late Professor M. W. Flinn, laid down three guiding principles. The books should be concerned with important fields of economic history; they should be surveys of the current state of scholarship rather than a vehicle for the specialist views of the authors, and above all, they were to be introductions to their subject and not 'a set of pre-packed conclusions'. These aims were admirably fulfilled by Professor Flinn and by his successor, Professor T. C. Smout, who took over the series in 1977. As it passes to its third editor and approaches its third decade, the principles remain the same.

Nevertheless, times change, even though principles do not. The series was launched when the study of economic history was burgeoning and new findings and fresh interpretations were threatening to overwhelm students — and sometimes their teachers. The series has expanded its scope, particularly in the area of social history — although the distinction between 'economic' and 'social' is sometimes hard to recognize and even more difficult to sustain. It has also extended geographically; its roots remain firmly British, but an increasing number of titles is concerned with the economic and social history of the wider world. However, some of the early titles can no longer claim to be introductions to the current state of scholarship; and the discipline as a whole lacks the heady growth of the 1960s and early 1970s. To overcome the first problem a number of new editions, or entirely new works, have been commissioned — some have already appeared. To deal with the second, the aim remains to publish up-to-date introductions to important areas of debate. If the series can demonstrate to students and their teachers the importance of the discipline of economic and social history and excite its further study, it will continue the task so ably begun by its first two editors.

The Queen's University of Belfast L. A. CLARKSON
General Editor

1 Introduction

The purposes of this book are threefold. The first is to introduce the reader to the burgeoning literature on England's agricultural regions in the sixteenth and seventeenth centuries, and to explain how historians construct them and arrive at different maps with varying boundaries. The second aim is to offer a simplified schedule of farming regions that will help the reader to perceive the main characteristics of England's landscape before venturing to absorb, compare and judge the detail that has been offered by specialist scholars. The third aim is to make more understandable the debate on the economic significance of agricultural innovations in the sixteenth and seventeenth centuries and their impact on these specialised regional economies. The list of innovations has not aroused controversy. But differences of opinion exist concerning their importance in improving agricultural output before 1750; did they amount to an agricultural revolution? An exposition of the differing viewpoints is offered here.

It is hoped that the discussion of these three topics will prove to be a helpful introduction to the voluminous literature on the agricultural history of the period 1500–1750, leaving the reader better informed to make his way alone hereafter. In preparing this book, I have borne in mind the needs not only of those whose main interest is the economic and social development of England as a whole, but also of those who prefer to work on a smaller canvas, and who wish to place their locality within a larger context. Above all, I hope to have made it easier for all readers to transpose the agrarian landscape of the sixteenth and seventeenth centuries from the mind's eye onto the visible landscape of the present, and so bring the past to life.

2 Agricultural Regions in General

In the days when Lord Ernle was writing an account of *English Farming Past and Present*, published in 1912, it was possible to write as though agriculture were one enterprise, moving first in one direction and then another, but possessing a recognisable homogeneity of aims and ends. One passage in Lord Ernle's text demonstrates the simplicity of this viewpoint. It purports to describe a general shift to pasture farming, followed by a general shift back to arable farming between the sixteenth and eighteenth centuries.

> Down to the middle of the reign of Elizabeth the enclosing and grazing movement continued. At subsequent intervals it renewed its special activity throughout the seventeenth century, when dairying began to claim a larger share of the attention of farmers. It was restrained or encouraged rather by natural causes than by legislation. Fluctuations in the prices of wool or corn, the increased profits of improved methods of arable farming, and the restoration of the fertility of the ancient tilled land, which was brought back to the plough after an enforced rest from excessive cropping, gradually restored the preponderance of tillage over pasture. [8, *61*].

Focusing his attention on the sixteenth century only, R.H. Tawney published in the same year as Lord Ernle *The Agrarian Problem in the Sixteenth Century*, in which many regional differences were made visible. But it was social, rather than agricultural, differences which concerned him, for he was tracing the fortunes of different classes of farmers, rather than their agricultural systems. Nevertheless, he was well aware of the agricultural differences that were implied in the varied local history of enclosure, finding a vigorous sixteenth-century movement in the Midlands, but not in the south-west, south-east or north. Moreover, Tawney used throughout his book manorial surveys from varied parts of the country; sharp regional differences in farming systems emerged from these documents [9, *212, 226, 106ff.*].

Tawney's book and later teaching were, therefore, extremely influential in directing subsequent scholars to the more detailed study of agrarian change in individual counties. Studies of Leicestershire, of Wiltshire and of Lincolnshire were post-war examples that owed much to his stimulus [40; 41; 44; 59]. They led on naturally to a more complete examination of the agrarian regions of England and Wales, resulting in 1967 in the publication of *The Agrarian History of England and Wales, vol. IV, 1500–1640* [22], and *The Agricultural Revolution* by Eric Kerridge [15]. Both contained similar, but not identical, maps of farming regions. A further analysis of farming regions, covering the period 1640–1750, appeared in 1985 in volume V of *The Agrarian History of England and Wales* [23].

A multiplicity of farming regions is now recognised in the early modern period, each having its own specialities and following its own distinctive path of economic and social change. It is no longer possible to write one account of the fortunes of English agriculture and rural society in the sixteenth and seventeenth centuries. This makes the study of the subject complicated, and perhaps daunting. But it presents a more lifelike and thus a more convincing picture, and it promises a better understanding of the processes of change. It allows us to envisage different farming systems working upon each other and modifying each other's practices while, at the same time, agriculture as a whole responded to the demands of industry and trade and fluctuating numbers of people to be fed. It also offers a challenge to local historians for they can always add more to our knowledge of the divergent experiences of different farming districts and different farming classes.

DEFINING AGRICULTURAL REGIONS

What is represented within the boundaries of an agricultural region? The way they are defined requires explanation first of all, since different scholars settle on different boundaries, and such variations are bewildering. But they have to be expected. Many factors are weighed and incorporated in one generalisation, and it is inevitable

that the relative importance of each factor should be judged differently by different authors, since most regions were undergoing
continuous change [16]. Every generalisation is a simplification and
hence, in some measure, it distorts the truth. But the different
judgements should not be on such a scale as to disguise totally a
central core of agreement. The principal factors that are considered
when the boundaries of farming regions are determined by historians are physical, agricultural, economic, social and political.

PHYSICAL FACTORS

Soils, subsoils, altitude, relief, rainfall – indeed climate generally –
place certain limitations on farmers' options. High altitudes and
above average rainfall favour grass-growing, whereas drier climates
favour arable farming. Hence the fundamental distinction between
highland and lowland England, which has always prevailed
throughout our history. The reasons for the contrast are clearly
and succinctly explained by Dudley Stamp in the following passage:

> to the north and west is Highland Britain, with most of the
> mountains and hill masses built up of old rocks resistant to
> weathering, and breaking down but slowly into soil. The rain-
> bearing winds come mainly from the west, so that ... rainfall is
> heavy, even excessive, with the result that grain crops fail to
> ripen.
>   Only the more sheltered valleys and the few plains ... are able
> to support human settlements ... Highland Britain is largely a
> rural area of discontinuous settlement, except on the lower drier
> parts.

Lowland Britain, on the other hand, is 'an undulating lowland of
varied aspect rather than a plain ... the hills rarely reach 1,000 feet
above sea level, and though the land varies in quality there are
relatively few parts which have not, or cannot, be cultivated.
Settlement has taken place over practically the whole: only a few
"islands" of poor land remain unfarmed' [20, 1–2]. Pastoral country
thus lies mainly in the north and west, arable country predominates
in the south and east. The line conventionally drawn between the
two zones runs from Teesmouth or Tynemouth in the north-east
to Weymouth or Exmouth in the south-west.

Such a division is useful as a first guide. But the drawing of a boundary immediately sets up a hazy border zone, where the two regions meet and the differences between them are blurred. Here is the first ambiguity. It is most conspicuous when differentiating the mainly pastoral from the mainly arable halves of England, for the dividing line runs across the central counties of the East and West Midlands. It means that both arable and pasture farming were possible here, and the preferences of farmers between the two choices were liable to shift under the influence of less permanent considerations than relief, climate and soil. Current produce prices, government policies and current weather trends could, and did, tilt the balance of farming from grass to grain and back again.

The soils of Britain vary in their composition and quality over very short distances. Within the larger contrast between highland and lowland England, therefore, refined analysis throws up many smaller farmer regions, which differed from the general complexion of those around them [20]. A prime distinction must be made between clay soils that themselves range from light to heavy, and calcareous soils of lime and chalk. Further local variations are introduced by other special features, such as the presence of peat layers in the fens of eastern England, and of saltmarshes along the coast, which were intermittently flooded by the tides. Different soils respond best to different uses. Heavy clay soils are difficult to cultivate, though not impossible. Hence, they will always grow grass, but in the right circumstances they will compensate for the labour of growing arable crops, particularly wheat and beans. Lighter clay soils are more easily worked, and so are well suited to arable farming. They will grow wheat and barley, beans and peas. The lightest soils which are calcareous, are more easily ploughed than any other, and offer themselves as arable lands first and foremost. But they are usually shallow and have to be regularly manured. Since they are found in hilly country, on wolds and downlands, an elegantly neat system for these soils was the sheep-corn regime, under which sheep were fed by day on the short, sweet grass of the hills, and folded by night on the fields, which they thereby manured so that barley and peas grew in abundance. This system is always described as an arable regime, though it could not have worked in the early modern period without the sheepflock [15, *42–5*].

Apart from the large intermediate zone of the central Midlands, where physical conditions have always permitted farmers to choose between a predominantly pastoral or predominantly arable system, smaller districts in both highland and lowland England can be identified which, on account of their physical conditions alone, departed in a marked way from the general practices prevailing around them. Thus sophisticated arable systems were practised in the sheltered valleys of the south-west, in the midst of a broader pastoral country. A celebrated example of highly professional arable cultivation was found in the Vale of Taunton Deane in Somerset [15, 115–16]. At the same period, an exceptional region in the south-east was the Weald of Kent, where the heavy clay soils imposed pasture farming, and woodland was still extensive, although it lay in a predominantly arable part of the kingdom [22, 57–9; 15; 132–3, 170–1]. In practice, then, many smaller farming regions can be identified within a larger whole, and even when the physical conditions are not strongly contrasted, as between chalklands (under a sheep-corn regime) and claylands (used for dairying) as in Wiltshire [44], farmers will exploit finer differences and combine choices in the use of land to make up their own distinctive mix. In pastoral country cattle-rearing may be preferred to dairying; pig-breeding may, or may not, be combined with dairying; horse-breeding may flourish alongside other pursuits in one district but not in another. The onlooker has to make many subjective judgements when classifying farming regions that are a complex dovetailing of several activities.

So far we have considered only the physical constraints on farming choices. Several others will be considered below. But in order to illustrate and further underline the problems of classification, it is worth noticing the methods used by the Ministry of Agriculture and Fisheries when in 1939 it prepared a *Types of Farming* map of England and Wales, published in 1941, depicting present-day regions. It had abundant statistical evidence, which historians always lack, drawn from the annual agricultural statistics of acreages, crops and stock, supplied by every farmer in the kingdom. With this information, it established three broad farming types, 'pastoral', 'intermediate' and 'arable', and added a fourth 'miscellaneous' category. Even then the mapmakers had to make an arbitrary judgement on what differentiated one type from another. It was decided that a farm with arable extending over more than two-thirds of its land belonged in the 'arable' category, one with between one-third and two-thirds of its land in arable was 'intermediate', and those with less than one-third of their land under the plough were 'pastoral'. It then subdivided the first three groups into seventeen farming subtypes, and a more careful scrutiny of secondary pursuits was made to classify farms into one of the seventeen subtypes. Nevertheless, the final decision was still far from easy. The explanatory text that was prepared to accompany the map emphasised the lack of complete uniformity that characterised almost all areas, and the 'non-uniformity' which dominated some [21]. This example of the difficulties in the way of generalisation, when abundant information is available, needs to be borne in mind when examining the various farming regions that have been defined by historians with far less information to hand.

SOCIAL FACTORS

While climate, topography and soils impose certain limitations on farmers, they never remove all freedom of choice. Farming systems are shaped by men, responding to many changing social, economic and political considerations, and in drawing boundaries the historian is especially concerned to take these into account. This is not a simple matter since farming was the main livelihood of, perhaps, two-thirds of the population, and within this large group were several social classes farming with different objectives. Many small farmers at the beginning of the sixteenth century, at least, farmed largely for subsistence. They had to produce something of everything in order to furnish the basic necessities for their households. Grain was essential for their bread, gruel and drink, and animals were indispensable to fertilise the arable fields, as well as providing meat, milk and other dairy produce. The location of their farms was not always ideal for all these purposes, but the physical constraints were overridden by compelling human needs.

The social structure of the community, as it existed at any given date, also constrained farming choices. Dairy farming, for example, was ideally suited in the sixteenth century to communities in which small family farmers occupied the land. They could manage a herd of 15 or 20 cows with the labour of their members and little more, and the financial returns from cheese and butter production came in regularly. This speciality suited small farmers much better than cattle-rearing and fattening, which was designed for farmers with larger resources, who could afford to wait for their returns. Rearing and fattening, then, was likely to be found in regions with more gentry and yeomen in the population. Yet again, grain farming is most economically carried out by farmers with large acreages and a labour force of hired workers who are readily available in the busy seasons. It was found in regions with gentry, yeomen and many wage labourers.

The social structure of different regions of the kingdom was already set in a certain mould in the sixteenth century, having been shaped by events in the remote, and not so remote, past. These existing contours of local society predisposed regions towards certain farming systems that might not have been chosen had the decision rested on agricultural considerations alone, or been shaped by current economic and political advantage. But they could be long-enduring characteristics, and extremely influential in determining agricultural choices. Nevertheless, being man-made, they were capable of being changed, not in a moment, of course, but within a generation.

An example of a fairly recent man-made change of the fifteenth century, which was alive in men's memories in the sixteenth century, was brought about by the desertion of some settlements after the Black Death. When shrunken populations in small hamlets moved elsewhere, the only reasonable use that could be made of the abandoned land, if it were not to revert to scrub and then woodland, was to let it down to grass and run sheep over it. A dramatic change in the farming regime occurred in such cases, precipitated by the dramatic change in the numbers of people living in the place. The former arable, grain and livestock economy had to be replaced by a pastoral one that required far fewer people to work it. The land was now managed by a shepherd or two and perhaps a warrener to guard a rabbit warren.

When less dramatic, but equally significant social changes followed in the sixteenth century, all reacted in some way on agricultural regimes. When landlords insisted on enclosure, for example, the increased value and rents of land could result in the disappearance of small farmers, and a significant increase in the number of medium and/or large farmers. This could lead, in the right circumstances, to a changing emphasis in arable farming, involving more grain-growing and less livestock-rearing. Yet again, the increasing population of the sixteenth century could force the division of some family farms into more smallholdings. In a pastoral region this might lead to more dairying and less stock-fattening. Any social change, in fact, was likely to precipitate some modification, however subtle, in farming specialities, because the relationship between the land and labour resources of communities and of individual households had been altered.

The historian who seeks to define agricultural regions soon becomes aware of the dovetailing of farming types with social structure, and when taking this factor into account often uses the term 'agrarian region' rather than 'agricultural region' to signal the difference. The example given above of the connexion between shrunken, or disappearing, village populations and sheep farming is the clearest and simplest case of that interrelationship; pasture farming could be managed with far fewer people than arable farming. But the class composition of rural communities also encouraged some more refined agricultural choices, such as dairying or hop-growing, rather than horse-breeding or cattle-raising. The subtleties of these interconnections await fuller exploration and explanation. They were certainly not concealed from contemporaries, as some of their discerning comments bear witness. A crisp, clear observation on this score was made by a parson in the Vale of White Horse, Berkshire, in the mid-eighteenth century. 'If there is one inconvenience that often attends a great venison county, it is generally overstocked with proud nobility and overbearing commoners; and as we are without these sore plagues in our neighbourhood, I can very well dispense with the rent of venison and rest contented with Ilsley mutton.' (1).

Economic Factors

Every farming region is an interlocking structure of agricultural and social elements, but that is not all. Both agriculture and society submit to change under the influence of outside forces. In a period of heightened market activity, such as the sixteenth and seventeenth centuries, economic pressures were strong and yet constantly shifting their ground. The strongest were generated by changes in the size of the population needing to be fed. High mortality at the

17

time of the Black Death in the fourteenth century had reduced the population by a third and hence reduced the demand for grain. Farmers had been encouraged instead to produce more wool to supply the flourishing cloth industry. In the sixteenth century, however, the tide turned again: in 1541 the population of England was 2.75 million, by 1601 it was 4.1 million, and by 1641 just over 5 million [4, i, 2]. Since so many more mouths had to be fed, grain shortages forced up its price, and farmers ploughed up pasture in order to put more land under arable once more [22, *Ch. ix*]. In the first half of the seventeenth century the rate of population growth slowed down, and since farmers for a hundred years at least had put great efforts into producing grain, they now faced a crisis of overproduction. After 1650 grain prices fell remorselessly, and once again farmers had to change their ways [23, *Ch. xiii*]. For those who did not farm the best wheat and barley land, meat and dairy produce, fruit, vegetables or other special crops became more profitable than grain or wool. Industrial populations in the Midlands and north, for example, demanded more cheese and butter, and local farmers in north Shropshire vigorously responded [34]. Farmers living near large towns perceived a rising demand for pork, as well as vegetables and fruit, and slanted their efforts in these new directions [23, ii, *337, 386, 503ff.*]. Royalist soldiers encamped in the West Midlands during the Civil War developed a taste for cider which encouraged farmers in Herefordshire and Worcestershire to improve its quality, and market it nationally [23, i, *190*; 23, ii, *303, 344–6*].

Means of transport was another changing factor in the situation, which in some places greatly improved farmers' opportunities to sell their produce in more distant markets. Many new stretches of river were cleared for boat traffic: for example, the Thames was gradually made fully navigable between London and Oxford between 1540 and 1635; and a variety of schemes for other rivers were put into effect in the seventeenth century, especially after 1660 [10, *349–51*]. Regular carrying services by road, using larger waggons, and by coastal vessel were greatly improved in the seventeenth century. In the 1630s farmers could ship their produce from the north Kent and Essex coasts twice a day to London, and from Maidenhead and Windsor boats went twice a week [10, *351–2*]. Such facilities could make all the difference between success and failure for farmers producing perishable foods like fruit and vegetables. In short, within the constraints of the agricultural and social frameworks of their lives, and changing economic circumstances, farmers continuously adapted their farming systems.

Political Factors

Some changing circumstances for farmers were politically contrived, as a result of government policies, though these were not necessarily devised with clear farming purposes in mind. The fall of wool prices in the seventeenth century, for example, was aggravated by the government's determination not to allow wool to be exported; exports, it was thought, would profit foreign textile industries at the expense of the English industry. In fact, the demand for English wool was falling because of the competition of Spanish wool, and lighter cloths made of linen and cotton [23, II, 363–6]. Cattle farmers were deeply affected by two Irish Cattle Acts, passed in 1663 and 1667, which forbade the import of young animals from Ireland. The Acts were intended to meet the grievances of landlords in the south and south-east, who complained loudly of the falling rents of their fattening pastures and blamed the import of Irish cattle for their woes. In fact, the rearers of young cattle in the north and west benefited most from this legislation; the shortage of young cattle formerly brought from Ireland, to be fattened ultimately in the south and east of England, gave them just the encouragement they needed to intensify their cattle-rearing activities. They prospered unexpectedly [23, II, 346–56; 29], and were launched on a new phase of economic growth.

Conclusion

Because so many different influences shaped and modified farming systems, classifying them in the sixteenth and seventeenth centuries is bound to be a somewhat subjective matter, which can never be reduced to a scientific system. So we return to answer our first question: what is represented within the boundaries of an agricultural region, and why do authors differ? The brief answer is that historians amalgamate as many different considerations as they perceive in their final verdict on the character and extent of distinctive farming regions. As J.D. Marshall phrases it, boundaries

are 'a matter of perception, usage, and convenience' [16, *15*].
Moreover, when farming specialities and systems are portrayed on
a map that purports to cover a century or more, it is an approximation, for it suppresses variety and casts a haze over the situation
on every geographical and chronological frontier. But every historical generalisation is an approximation. Its value lies in clarifying
the direction of large changes, and encouraging further investigation
of the small ones.

A long answer to the question how historians define farming
regions could carry us into a discussion of the varied documents
now at their disposal. These have become far more abundant in the
last fifty years and so have enhanced the possibilities of portraying
regions more accurately. This is the reason for the many attempts
at defining regions that have been made in recent years, in contrast
with the past, and for the lively debate, especially among geographers, about what a region is. Such a discussion must be cut short
here. Suffice it to say that present-day historians and geographers
strive to incorporate in their map-making more detailed information
than was available hitherto, using estate and farm accounts, descrip-
tive statements and above all probate inventories. These last list
the farm goods of men and women at death. If the crops, stock
and equipment derive from a fully working farm, and best of all,
one at work in the summer season, the inventory will give a fair
idea of its owner's main farming pursuits. Adding a sufficient
number of these together will sum up the farming specialities and
so help to define the extent of a region.

In practice, the use of these inventories raises further problems
to tax the specialist. They need only be briefly mentioned here,
but they could slant the result in different directions. The inventories that survive do not faithfully reflect the class composition of
society. Yeomen and prosperous husbandmen are better represented than small husbandmen and cottagers. If the farming
specialities of a region are defined as the market specialities, then
this is not a matter of any moment, since gentlemen, yeomen and
well-to-do husbandmen – the well-represented classes – offered
more to the market than the rest. With this assurance historians
have satisfactorily used samples of inventories to identify the main
pursuits of regions. The use of the computer now makes it possible
to analyse much larger numbers of inventories, and we shall see
much more comprehensive analyses in the future [83]. They will

multiply our examples, but the refining of our perceptions must also be expected, since they are likely to incorporate more information from the inventories of cottagers and labourers. These will reflect in paler colours the market specialities of regions, but will probably give a more accurate impression of the objectives of the majority of the rural population, since cottagers and labourers were more numerous than gentlemen or yeomen.

A good illustration of this can be taken from the expanding dairying region of the north Shropshire plain in the seventeenth century. The inventories of ordinary men show a dairying interest, in the number of cows owned and the presence of dairying equipment. But the mean number of cows per farmer between 1600 and 1670 was 4.9, not an impressive figure [34, *188*, recalculating Table 2]. The commercial scale of the business shows best in the inventories of the larger men, notably John Curdworth of Adderley with 37 cows, 2 bulls, 2 heifers, and 12 weaning calves in 1624, and Richard Furber, yeoman, of Shavington in the same parish in 1660 with 62 cows and 77 other cattle of varying ages and sexes [34, *177–8*].

Whatever the practical difficulties of defining and refining regions, the task is made worth while by the often unexpected contrasts between them that are so revealed. These draw attention to features that may, in the end, explain processes of change more fully than before. At the very least, regional analysis points to features worthy of further investigation. New insights have already been gained from this kind of work that have broadened our understanding of the progress of enclosure [22, *240–55*] and of the location of rural industries [24]. It has modified conventional explanations for the political and religious loyalties of different regions in the Civil War [22, *111–12, 462–5*; 25]. It has helped to explain the attitudes of the Clubmen: in south-western England those sympathising with the royalists have been shown to have come overwhelmingly from chalk downland country, while those sympathising with Parliament came from woodland and pastoral regions [64]. Such connexions between farming types, social structure and political and religious affiliations do not, however, go as far as some that were alleged by people actually living in the seventeenth century. In the 1680s, for example, John Aubrey described the inhabitants of the cheese country of north Wiltshire and the Vale of Gloucester as:

21

phlegmatic, skins pale and livid, slow and dull, heavy of spirit... they only milk cows and make cheese: they feed chiefly on milk meats, which cool their brains too much and hurts their inventions. These circumstances make them melancholy, contemplative, and malicious; by consequence thereof come more lawsuits out of north Wiltshire, at least double the number to the southern parts. And by the same reason, they are generally more apt to be fanatics ... In Malmesbury Hundred, etc. (the wet clayey parts) there have even been reputed witches.

In the sheep-corn country of the Wiltshire chalklands, on the other hand, 'where 'tis all upon tillage, and where the shepherds labour hard, their flesh is hard, their bodies strong. Being weary after hard labour, they have no leisure to read or contemplate religion, but go to bed to their rest to rise betimes the next morning to their labour'. (2) Regional analysis is a key that may unlock many doors.

3 Some Regional Classifications Illustrated and Compared

In this chapter some examples of regional analysis at different levels are presented. They are practical illustrations to show the varied features of an agrarian region being displayed by scholars at different scales of magnification. The main factors – physical, social, economic and political – that shaped their agrarian structure have been enumerated above. But according to the size of the region chosen for study, attention is concentrated more on some features than on others.

Some scholars have preferred the insights to be gained from examining large regions, others have chosen to study small regions in a degree of detail that reaches down to a close examination of individual parishes, others have selected regions of intermediate size. In consequence, different features come to prominence and the resulting maps differ because of the varying emphasis that scholars choose to put on each item.

An example of regions defined primarily by their agricultural methods is given in *The Agricultural Revolution* by Professor Kerridge. In determining the boundaries and names of regions, it follows the lead of the late-eighteenth century agricultural writers, especially William Marshall, who also put techniques first. Some very large farming regions emerge as a result, as well as some very small. Kerridge's 'farming countries', however, are not always the home of one farming system, for he defines them differently. The 'farming countries' are areas in which he detects 'uniformity or similarity in a single plan, or in two or more competing plans, of farm management, sufficient to permit the formulation of general conclusions on husbandry systems' [15, *41*]. The agricultural regimes so described lasted throughout the period 1500–1800, subject, of course, to changes that resulted from innovations to which the second half of his book is devoted.

One of the largest of Kerridge's regions is the Midland Plain,

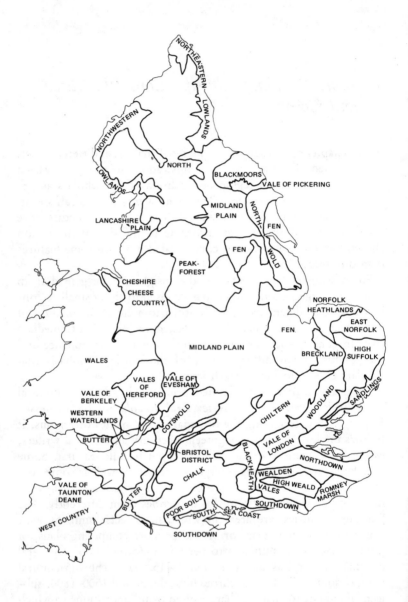

Map I. English farming countries in the age of the agricultural revolution

which stretches from Shropshire to the Vale of York and north Lincolnshire, and from Wiltshire to the Vale of Aylesbury in Buckinghamshire. This region thus encompasses many vales of clay soils, some uplands like the limestone cliff and heath of Lincolnshire, and forests like the three stretching down the eastern side of Northamptonshire, as well as Galtres and Knaresborough forests in Yorkshire, and Needwood forest in Staffordshire [15, *91–113*]. Professor Kerridge unifies the region by calling 'the country as a whole' a 'disjointed rolling plain', and among its common features stresses its common field husbandry, at the beginning of the period at least, based on a three-course rotation, though two- and four-course rotations were also found. The principal objective throughout the whole plain was grain-growing, though some dairying and cattle-feeding were also carried on. A larger interest in dairying for cheese is acknowledged towards the north and west of the Midland plain, and for butter towards the south and east; specialised dairying is also allowed for in other areas embedded in the plain, along the river valleys of the Soar, the Trent and the Dove. Yet other exceptional regions are recognised around Leicester, Rugby, Market Harborough and other places specialising in stock-fattening, and the routines of these graziers are described in some detail. Because Professor Kerridge allows in his definitions of a 'farming country' for 'two or more competing plans', and since forest economies, dairying regions, cattle-feeding and grain-growing countries are all encompassed in the Midland Plain, it is clear that other investigators would have no difficulty in breaking up this large area into smaller regions, whereby they would arrive at greater homogeneity in the farming systems of each. Such regions classified by agricultural criteria alone would be further refined if attention were also paid to their social structure.

Another of the large regions defined by Professor Kerridge is the fen. Its boundaries embrace the fens around the Wash, the whole of the Lincolnshire marshland coast, and the whole of east Yorkshire, namely the part usually known as Holderness. The unifying characteristics chosen here are the fact that it is entirely an area of water-formed lands, whether of silt on the coast or peat on a clay base inland, and it is everywhere divided by dykes and drains. The author stresses that grazing was abundant in summer, and much intercommoning was practised between townships. Fishing and fowling were important activities, and reeds, sedge and rushes were significant resources for basket-making, litter, thatch and fuel. Nevertheless, farming conditions in the fens or marshes are described as 'highly variable', and the stable element in the rural economy is deemed to be 'the cultivated uplands where the farmers lived and where husbandry practices, though greatly influenced by the proximity of fens, was [sic] in some respects assimilated to those of the neighbouring countries.' The crop rotations varied from four– to three- to two-course and 'every year's ground' [15, *138–42*]. In other words, this area contained much variety in detail in its husbandry practices, and could be subdivided into many smaller farming regions. Short of this, a division could be justified between the fens around the Wash, which had far more land flooded by rivers than the coastal marshland, and the marshland along the Lincolnshire coast and in the Vale of Holderness where river-flooded fen was small and much more pasture represented gains from the sea. Such a division would enable one distinction, at least, to be better emphasised, namely that between areas with common fields (in Holderness and the Lincolnshire marshlands), and the fenlands where there were none. It would also be possible, as Professor Kerridge implies, to distinguish between those parishes firmly embedded in the deep fen as in Cambridgeshire, where the farming was not dependent on the use of land outside the region, and those in the marsh which were associated with the upland systems of the neighbouring wolds of Yorkshire and Lincolnshire; some wold farmers worked their wold land in conjunction with grazings in the coastal marshland.

Kerridge's portrayal of farming regions represents the situation in 1800 more faithfully than in 1500, relying as it does on the regional classification of observers who saw the countryside at the later date. Kerridge projects their regional classification, with certain modifications, back over the previous three hundred years. It should again be stressed that farming techniques are the main guide to this judgement [17].

Another delineation of farming regions is contained in *The Agrarian History of England and Wales*, vols. IV and V,I [22; 23], where the definition of regions is guided mainly by documentary information drawn from the two separate periods concerned, 1500–1640 in vol. IV [22], 1640–1750 in vol. V [23, I]. Some of the regions arrived at are smaller and some are larger than those in *The Agricultural Revolution*, but a more varied mixture of elements is taken into consideration in defining them. As the title of the work indicates, it surveys 'agrarian' rather than purely 'agricultural' features. The commercial objectives of farming, that is the farming specialities, are accommodated first, and changes that occurred between the two periods are underlined. But attempts are also made to take account of other features of local economies, especially rural industries which dovetailed with farming in the rural household, and attention is given, in broad terms, to the differing social structures of regions.

As a result of this different viewpoint, the Midland Plain, which is an extensive area in Professor Kerridge's classification, is divided in *The Agrarian History, vol.* IV, into several smaller units. These are: (i) a large area of vale land where mixed (meaning more arable-biased) farming combined corn and stock production (3); (ii) an upland area of limestone cliff and heath (north and south of Lincoln) where arable-based farming was directed at grain and sheep; and (iii) a number of woodpasture areas such as the Northamptonshire, Charnwood and Leicester forests, where stock-rearing and pig-keeping went hand in hand with horse-breeding, and Galtres forest (Yorks.) where stock-fattening and pig-keeping were the specialities [22, *38–40, 35–6*]. The differing class structure of vale and forest regions is emphasised in the accompanying text, and is still more cogently argued by Professor Alan Everitt in Chapter VIII, writing on farm labourers, where their greater numbers and greater agricultural resources in the commons and wastes of the forests are underlined, in contrast with the situation in the vales, where the commons were limited, the land was already taken up in larger units than mere smallholdings, and the poor could find no refuge [22, *28ff., 396ff., 409ff.*].

In *The Agrarian History, vol.* IV, the fenland region of eastern England, in contrast with the much larger area selected by Professor Kerridge, is confined to the fens around the Wash and one other distinct area further north, covering the Isle of Axholme and Hatfield Chase. Both regions of fen are classified as pastoral, specialising in stock-fattening with horse-breeding, dairying, fishing and (in the watery parts) fowling. A principal feature of the social structure of the fenland is also underlined, namely its many small peasants and few gentry. But the difficulties of classification when change was continuously under way are well demonstrated in this case. The fens though pastoral in the past were undergoing

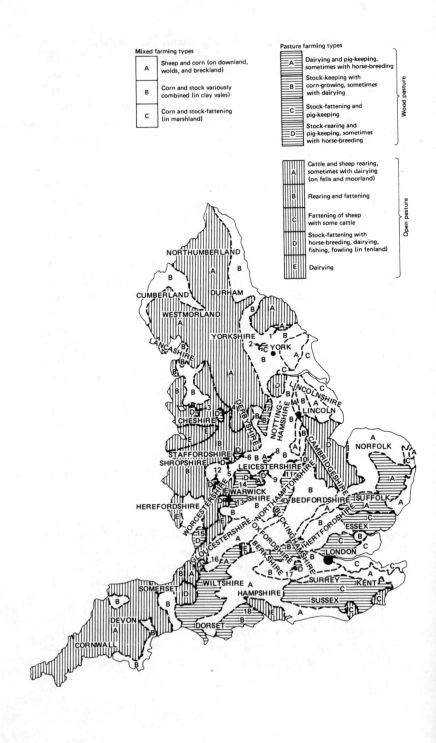

Mixed farming types

A Sheep and corn (on downland, wolds, and breckland)

B Corn and stock variously combined (in clay vales)

C Corn and stock-fattening (in marshland)

Pasture farming types

A Dairying and pig-keeping, sometimes with horse-breeding

B Stock-keeping with corn-growing, sometimes with dairying

C Stock-fattening and pig-keeping

D Stock-rearing and pig-keeping, sometimes with horse-breeding

Wood pasture

A Cattle and sheep rearing, sometimes with dairying (on fells and moorland)

B Rearing and fattening

C Fattening of sheep with some cattle

D Stock-fattening with horse-breeding, dairying, fishing, fowling (in fenland)

E Dairying

Open pasture

Map II. Farming regions in England 1500–1640

This is a very tentative map of farming regions in England in the sixteenth and early seventeenth centuries. The boundaries between the regions are the most tentative of all, and will certainly require amendment in the light of more detailed local investigation. Not all royal forests are indicated on the map since it has not been possible to identify all their boundaries with certainty. All, however, belong in the category of wood-pasture regions, and most lie within or on the fringe of regions of open pasture. The Warwickshire Arden is depicted as a wood-pasture region undergoing change from stock-rearing, etc., (D) to dairying etc. (A). The forests are numbered as follows: 1, Pickering Forest; 2, Galtres Forest; 3, Delamere Forest; 4, Macclesfield Forest; 5, Cannock Forest; 6, Needwood Forest; 7, Sherwood Forest; 8, Charnwood Forest; 9, Leicester Forest; 10, Leighfield or Rutland Forest; 11, Rockingham, Whittlewood, and Salcey Forests, Northamptonshire; 12, Kinver Forest; 13, Feckenham Forest; 14, Forest of Arden, 15, Forest of Dean; 16, Kingswood Forest; 17, Windsor Forest; 18, New Forest.

drainage in the seventeenth century, and the land which was successfully drained was usually put under the plough. It is impossible to estimate the acreages so transformed, especially since much drainage was undone in the Civil War. The whole region, in fact, was in the midst of a long, slow transformation from a pastoral to an arable region.

The marshland coast of Lincolnshire, and Holderness in Yorkshire, together with the Ouse–Humber marshes, which are included in Professor Kerridge's 'fenland', are described in both volumes of *The Agrarian History*, as arable-biased farming regions. They were engaged in both grain-growing and stock-fattening though, as is illustrated below, stock-fattening later gained ground [22, *35–6*; 23, I, *76–8*].

In another part of the country, in the Weald of Kent, comparison of Kerridge's and Thirsk's classifications shows differences on the map, but both texts contain an agreed core of description that covers the essential features. In *The Agrarian History, vol.* IV, the High and the Low Wealds are amalgamated in one region, whereas in Professor Kerridge's account they are separated. *The Agrarian History* classifies the whole area as a woodpasture country, given over to stock-fattening and pig-keeping, though the text makes clear that heath and coppice were more conspicuous in the High Weald than the Low, and that stock-fattening was probably taking over from an older, medieval, interest in dairying. The existence of by-employments in ironworking and in cloth-making are judged important in explaining the large populations and small size of arable holdings in the region. Professor Kerridge separates the High Weald from the Low Weald, since their soils were different. He notes as one common feature of both that neither had common fields. But he differentiates them by emphasising in the Low Weald its cold clays, and its up-and-down husbandry, whereby grassland was ploughed up for two to four years, followed by six to seven years of grass. The main business was cattle-fattening with some sheep-fattening, the store cattle and sheep being brought in from the High Weald as well as elsewhere. The High Weald had much woodland and rough grazing, some very poor, almost worthless, arable land, and many rabbit warrens. Its main interest was in breeding young cattle for the Low Weald farmers among others, and grazing sheep for the benefit to the grassland of mixed grazing by both kinds of animal. This description shows the economies of High and Low Weald closely interlocked [22, *57–9*; 15, *132–3, 170*]. In volume v of *The Agrarian History*, describing the period 1640–1750 [23, I, *270–3*], the Low Weald is separated from the High Weald and lowland heath, but not along exactly the same lines as in Professor Kerridge's map, for the 'lowland heath' runs on to the coast to take in part of the Kentish marsh. The reason for this variation lies in a significant change in the occupation of the marsh in the seventeenth century, which brought it into closer relationship with the Weald. Increasingly, High Wealden farmers used Romney marsh pastures as grazings, as a result of which parishes on the marsh saw their resident population much depleted. Here we encounter the problems of defining boundaries between regions when farmers themselves overstepped them. A more ample description is also given in vol. v, compared with vol. IV, of the varied activities pursued in the High Weald and lowland heath. They might consist in subsistence corn-growing with cattle-rearing or dairying and/or grazing, but they might also include industries as by-employments. A new phenomenon of the period was the increasing number of orchards and hop gardens – fruit, in the eastern High Weald especially, and hops in the eastern and central portion of the High Weald and lowland heath. Nevertheless, in all three accounts of the Kentish Weald, the essential facts are in accord with one another, differences arising from an inclination

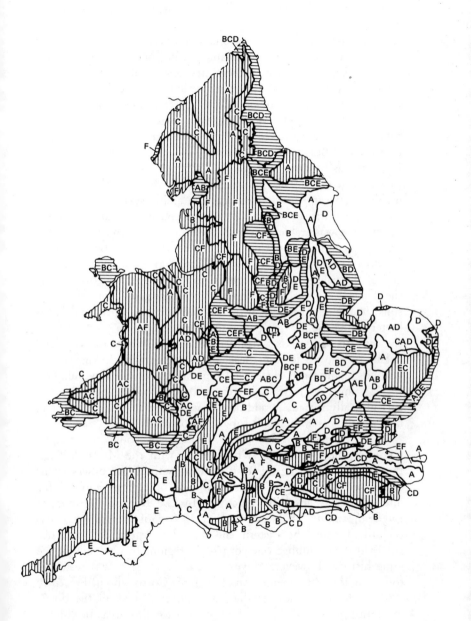

Map III. Farming regions of England and Wales, 1640–1750

sometimes to link, and sometimes not to link, interdependent regions and to ignore lesser differences in land use and farming objectives.

Two final examples of regional classification concern smaller districts investigated in considerable detail. They show the work of local historians in refining generalities, and clarifying the complex interactions of social and farming structures on each other. The first concerns the vale of the White Horse in Berkshire, a mainly clay vale of light and heavy loams.(4) In some parts of England such a vale could be expected to show the same agricultural specialisation throughout. But market forces were strong in Berkshire because London was so close, and the land was sufficiently flexible in use to be turned either to pasture or arable. In fact, the Vale was divided between separate pastoral and arable systems. A pastoral dairying country occupied the western end of the vale, while the eastern end was arable. But even then the eastern end was not one uniform region. In his survey of the agriculture of Berkshire for the Board of Agriculture (1809), William Mavor divided the eastern end of the vale into three regions. In the northern section, lining the Thames, more barley and turnips were grown on its lighter soils; in the vale proper, in the middle portion, wheat and beans better suited the heavier loams; the third belt to the south of these two was recognised as a frontier zone because here the vale land begins to rise up to join the chalk of the downs. Its agriculture was not significantly different from the rest of the clay vale [23, I, 326], and yet it was differentiated socially (see below). And agriculturally it had, in addition to its lighter loams, a greensand belt which was well suited for use as orchards.

In *The Agricultural Revolution*, Professor Kerridge included the whole of the vale of the White Horse with the Midland Plain, thereby amalgamating this clay lowland with all the others in his 'disjointed rolling plain'. His definition of regions allowed for internal variety in their objectives. In *The Agrarian History of England, vol.* IV, the western dairying end of the vale was distinguished from the eastern end and the eastern end was not subdivided but dubbed one corn/stock region. In volume V of the same history, however, Mavor's tripartite division was restored, following the recommendation of Dr Janie Cottis who had recently made a detailed study of the Vale and examined all the documentation now available to the historian for this area. In volume v Dr Wordie explains the broad similarity of farming systems in all three arable districts [23, 1, *337–9*]. But they exhibited social differences which affected farming change. The relationship between manors and parishes was notably different as between the northern end and the rising scarpland at the southern end of the Vale. In the north, one manor usually occupied the whole of one parish. Lordly authority was thus significantly concentrated. As a result, easier and earlier enclosure was possible, since fewer land-holders had to be brought into agreement. In the south, by contrast, parishes usually had more than one manor in them, and freeholders were twice as numerous. Almost certainly these two characteristics are interlinked, since divided lordship in a parish was liable to weaken the will and ability of individual landowners to extinguish freeholds. Certainly, judging by the results, small farmers with 20–100 acres apiece found it easier to hold their own in the southern district and hence they slowed down the rate of enclosure.

It is also worth noting in this southern district yet another dovetailing of social and agricultural features: fruit-growing and hop-growing suited its greensand soils, but they also suited the working regimes of this class of farmers. Family labour was likely to be sufficient to deal with these profitable sidelines in all except the busiest harvesting season, and (for hops) in the bine-tying season. Thus social leanings and agricultural possibilities matched each other.

While at one level of investigation differences in the agrarian development of these three belts of vale land are perceptible, at another their experiences are plainly distinguishable as a group from those of other clay vales in the Midlands in the period 1640–1750. In other places, when grain prices were falling and labour costs rising, farmers showed a tendency to turn the heavier wheat and bean land to pasture. They did not do so here, probably because excellent river transport along the Thames to London gave them an assured market for their grain, and Abingdon was a busy and reliable river port and malting centre. All these three arable belts therefore remained notable for their small farms and surviving common fields, in contrast with other claylands of England, where peasants with small holdings found it difficult to survive the unfavourable economic conditions, and historians have written of the disappearance of the small landowner.

Certain distinctive features of the dairying country at the western end of the Vale of the White Horse also deserve to be underlined, for its experiences did not altogether coincide with other dairying countries elsewhere. Usually in the seventeenth century dairying regions were inhabited by many small farmers – owner-occupiers, very often – who were not able to afford such luxuries as floated water meadows to provide them with early spring grass. In the vale of the White Horse they did, possibly because they were small tenants under great landowners and benefited from the practical expertise and demonstrations of the efficacy of water meadows by their landlords, who were masters of the art on the nearby Wiltshire downlands. Nor did they rely entirely on permanent grassland, but showed an interest in experiments with clover and ryegrass as fodder crops in arable rotations, again, perhaps, profiting by observations of farming on the downs. A continuing interest in arable fodder crops might gradually have moved the region away from pastoral dairying towards a system better described as 'mixed corn and grass with a substantial dairying side'. But in the end, the region did not change its ways so radically. It turned against ploughing and temporary pastures and preferred permanent grassland. Thus the western end of the Vale of the White Horse remained a dairying area, even though it underwent some changes between 1660 and 1750 that were not common to all dairying country.

The second study to be used as an illustration of detailed regional variation concerns three Lincolnshire marshland parishes, Bradley, Scartho and Humberston, between 1520 and 1730.(5) It shows their similar agricultural potential being put to use by communities with differing social structures and a different history of land-ownership. Agrarian change, as a result, moved along at a different pace in each case. The same external pressures can be seen affecting them all, but one parish was dramatically transformed 100 years before the second changed similarly, and the third managed to resist change the longest, though by 1730 it too was drifting along the same path. The social influences on farming specialities are well illustrated here.

Agriculturally all three parishes were much alike, having clay soils that permitted arable farming – on a two-course rotation in common fields – and generous grazing grounds, partly on drier pastures, the rest on wetter outmarsh. Their main commercial specialities were grain, cattle and sheep, the cattle being directed towards the dairy at first but then increasingly towards rearing and fattening for the butcher. Since livestock prices in the seventeenth century held up better than grain, the demand for grazing pastures built up steadily, and manifested itself in the marshland in the conversion of more and more arable land to pasture and the buying or leasing of its land by outsiders, particularly arable farmers from the nearby wolds. As a result, the agricultural balance tilted from an arable economy to a pastoral one.

Bradley succumbed first. It had only two manors, which partly through inheritance but principally through purchase, came into the hands of a determined individual with a policy for buying up the parish which he resolutely implemented. He did not acquire every acre, but he became the dominant owner, and by 1634 had enclosed and turned all his land to meadow and pasture, which he either used himself or rented (without buildings) to farmers from other parishes. Bradley had become what is called a 'closed' parish, dominated by one lord, who had the power to control its size and structure because he controlled so many of its farms.(6) Land in Humberston, on the other hand, was held of seven separate lordships whose owners were mostly absentees. These estates underwent so many changes of ownership in the sixteenth century that tenants went their way undisturbed. Here was a happy hunting ground for the middling farmer who wanted to lease, buy or sell pieces of land, and had the means to make full use of his common rights. These last were generous, since the parish had over 1000 acres of common pasture, as well as a progressively expanding area of saltmarsh grazing beyond the seabanks. Small farmers and owner-occupiers could survive in this environment, but since they found themselves increasingly in competition with more determined buyers, from within the parish as well as from outside, some of their plans were frustrated. Their numbers dwindled until, in 1704, an aggressive buyer appeared who bought all the land in sight. Humberston now went the way of Bradley; it became a 'closed' parish, and the owner prepared for enclosure. The third parish, Scartho, was full of people with small farms and with a manorial lord holding neither demesne nor customary land in the parish. Resident freeholders ruled the roost, and none showed strong acquisitive instincts. They bought and sold on a small scale, and were as inclined to divide their farms among their male children as to keep them whole for one heir. But even this parish from the end of the seventeenth century, 1690–1730, began to lose some of its small owner-occupiers and to surrender grazing land to farmers from other places. However, it was a slower process than that effected in a matter of eighteen years, 1582–1600, at Bradley, and in the early eighteenth century, 1704–8, at Humberston.

Certain generalisations can be filtered from the diverse experiences of these three parishes, showing the interdependence of social and agricultural developments. The farming regime gradually changed from a mixed arable to a pastoral one. This went hand in hand with a thinning of the ranks of small to medium owner-occupiers, and in two cases a resident gentleman came to dominate the scene, thereby creating two closed parishes. The hiring or selling of grazing land to outsiders increased, and the rate of this change was closely connected with the social changes already mentioned. Above all, agricultural specialisation in favour of livestock-rearing and fattening was intensified. This trend was encouraged by general price movements of grain and livestock, but it was greatly facilitated by accompanying social change.

4 Types of English Landscape and Farming Regions: a Simplified Schedule

As we have seen, it is possible to identify large or small agricultural regions according to the precision with which the many varied farming features contained in them are isolated, and according to the amount of attention paid to social structures, to economic and to other influences. Much depends also on the precise historical date to which any study pertains; the Lincolnshire parishes described above showed one parish a hundred years ahead of another in its shift to a pastoral system. All definitions of agricultural regions, however, must first pay regard to the constraints of the physical environment, and so these appear in the forefront of every classification. In this chapter, a schedule of regional farming types is presented, which simplifies and summarises existing knowledge for the sixteenth and seventeenth centuries, and should serve as an introduction to the more elaborate, more local, or sometimes more idiosyncratic, regions that are, or may be, devised in specialist studies.

Simplification can first be achieved by reducing the number of regions. Wisely ignoring all county boundaries, Professor Kerridge nevertheless arrived at forty-one farming 'countries' in England and Wales. In The Agrarian History [22] counties are grouped, and several farming regions identified in each group. The map finally places all types in one of three main classes: (i) 'mixed', meaning arable farming with the necessary livestock; (ii) pastoral in forests; and (iii) pastoral, in open pasture, including moorland. Each of the three groups is then subdivided, making twelve types in all. In vol. v [23, 1], arable systems are now shown to be more conspicuously varied than the pastoral, and so these are divided into two, the one more heavily arable in emphasis than the other – (i) arable and (ii) intermediate – while all pastoral systems are assembled in the third category. These three groups are each divided into six subgroups, making eighteen in all. But since the period

37

1640–1750 shows much greater variety in farming systems, particularly in the southern half of England, the labelling of the regions is far from simple; on the map some regions bear two or even three alternative labels.

As an introduction to this regional complexity, a simpler scheme has been suggested by Professor Alan Everitt in an essay that principally discusses the evolution of regions from the time of their first settlement. But it is equally useful here, since it reduces the number of regions, gives them names which readily convey the appearance of the landscape, and enables anyone having some familiarity with the physical structure of England to locate them easily on the map. An eightfold regional division of the kingdom is suggested, as follows: (i) downland; (ii) wold; (iii) fielden or champion areas, which we shall call vale lands, and divide between arable (fielden or champion) vales and pastoral vales; (iv) marshlands; (v) heathlands; (vi) forest (sometimes called woodpasture) areas; (vii) fell or moorland; and (viii) fenland [12, 16]. Graingrowing regions were the downland and wold areas, the emphatically pastoral were the forest and moorland areas. At both these extremes of the spectrum some changes in land use were in progress, though they were muted simply because these arable and pastoral regions had the natural attributes, and hence the best economic reasons, for continuing to grow grain and grass respectively. Those which most obviously vacillated between arable and pastoral systems in our period were: (i) the vales (depending on whether they lay in the drier eastern half or wetter western half of England (see above, p. 12–13), and depending on current prices for different farm products); (ii) the fenlands (because draining on a grander scale began in this period, and former grassland could then be ploughed); (iii) the marshlands (for several reasons, but mainly because of changing price relationships between grain and livestock); and (iv) the heathlands (because of changing opportunities for improving arable cultivation through the introduction of new crops like buckwheat, clover and sainfoin).

WOLDS AND DOWNLAND

Because of their early settlement history, the wolds and downlands are separately considered in Everitt's categorisation, but may be

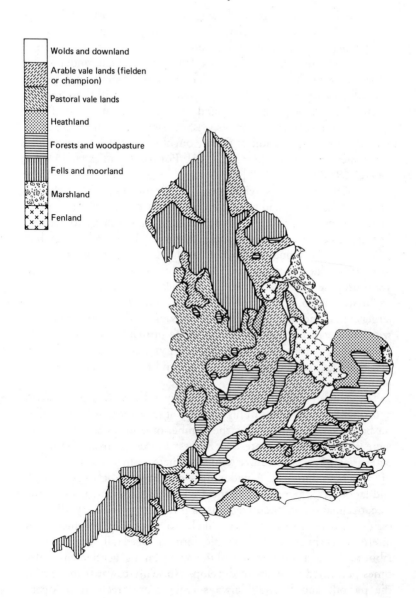

Legend:

- Wolds and downland
- Arable vale lands (fielden or champion)
- Pastoral vale lands
- Heathland
- Forests and woodpasture
- Fells and moorland
- Marshland
- Fenland

Map IV. Farming regions: a simplified schedule

amalgamated in the sixteenth and seventeenth centuries because their development was similar. Both types of region produced some of the best grain, mostly in common fields, cultivated under a three-course rotation. Theirs was a sheep-corn system, under which large sheep flocks fertilised the shallow soils at night and fed on the hill pastures by day.(7) In the period up to 1650 these regions profited by high grain prices and at first moderately good, but then discouraging, wool prices.(8) Some enclosure took place, as well as the amalgamation of farms, but throughout grain remained the principal objective [22; 23, 1; 15]. For local studies see [44; 56; 58–92; 59, 79–91, 159–78].

Agrarian conditions reinforced the advantages of natural conditions. Most down and wold country had ready access to coastwise or river transport, and grain could thus be carried easily to market. For example, the north and south downs of Kent and Sussex had water transport along the Thames or through the English Channel to London; the Lincolnshire and Yorkshire wolds had an east coast sea route; the Cotswolds used the Thames to London or the Severn to Bristol. Also, grain-growing is more efficiently carried on in larger units, and the social structure of wold and downland villages was already inclined to promote change in that direction. Usually village society took the shape of a tall pyramid with a wide base; often one gentleman was lord over all, and landless labourers were numerous.

The gentleman had the power to expand his own home farm and/or enlarge the farms of his tenantry at the expense of the smaller husbandmen, thus increasing the number of landless labourers who were available to cultivate and harvest the crops from the larger enterprises. Farms therefore tended to increase in size and numbers of farmers to diminish. A widening gap opened up between farmers and labourers, as middling farmers were squeezed out. Usually the sole occupation in these regions was farming; industry did not intrude since gentlemen were likely to oppose it in their own interests; at busy seasons arable farming used all the available labour. But a more humane philosophy among gentlemen sometimes permitted industry to develop, especially towards the end of the period, and it could always enter more freely into 'open' parishes, where no one lord held sway.

After 1650 this type of arable country was the least hard hit by falling grain prices since its grain was of the highest quality and continued to find markets, though some were different from those of the sixteenth century. From East Anglia much barley went as malt to Holland; from the Kentish downs it went to London brewers who passed on the waste to pig-fatteners. It is probable, though unprovable, that increased production compensated for lower prices. Cattle production entered as an additional item into farmers' activities, especially when they introduced artificial grasses (i.e. grass substitutes – clover, sainfoin and lucerne, etc.) into their arable rotations, and had better fodder to offer their animals. But unfavourable grain prices depressed the smaller farmers, depleted their numbers and widened the gap between the well-to-do farmers and the poor dependent labourers.

Vale Lands

(i) *Arable (Fielden or Champion) Country*

Arable vale lands form a category of agrarian region that was prominent in the East Midlands, in Hertfordshire, Buckinghamshire, Berkshire, and included the Vale of York and smaller areas in the north-west and south-west (see map IV). Settlement was based on village communities, with land farmed in common fields. But, agriculturally, vale lands were more capable than the wolds or downlands of tilting from arable to pastoral products as prices urged them. Vale soils varied from highly workable loams to difficult clays, but the shift towards pasture did not usually affect the best loams, only the heavy clays. Although much of this land still lay in common fields a great deal was enclosed in the sixteenth century, with the result that in some areas the new closes were converted to permanent pasture, while in others a system of convertible husbandry was adopted. Under this last regime the plough moved round the farm turning up leys that had been under grass for some years in order to grow crops for a spell; the land was then put back to grass for another few years, to recover its fertility again.(9)

A mixed corn-grass system was the result, in which cattle and sheep were used to fertilise the fields and the farmer's income was derived from both grain and animal sales. Some examples have been found of sixteenth-century gentlemen who turned determinedly to cattle- or sheep-keeping after enclosure, reducing their grain-growing to the very minimum and putting most of their land under grass [35, *14–20*]. By doing this, they were not necessarily choosing to produce items that were then increasing most rapidly in price, but rather saving themselves the trouble of employing and super- vising large numbers of labourers. Economic trends in the later seventeenth century, however, pushed farmers more firmly in a pastoral direction, since wages began to rise towards the end, and were the cause of much complaint in the early eighteenth century. In short, the change to livestock farming was accelerated in all clay vales after about 1650 by falling grain prices, helped by continuing enclosure. Dairying expanded in some places, fattening expanded in others [59, *92–107, 179–96*]. Fodder was procured by the grow- ing of more legumes on the arable fields – clover, ryegrass, trefoil, lucerne, sainfoin [38]. But some farmers made a return to the solutions of the fifteenth century; they turned these convertible lands back to permanent pasture [68].

Where the same amount of arable land continued to be cultivated as before, new crops were introduced into the rotations, which supplemented the conventional ones and paid better. Woad was used to break up old pasture; new fodder crops like clover, sainfoin and lucerne were introduced in place of the more labour-demanding beans and peas to feed livestock and maintain fertility.

(ii) *Pastoral Vales*

The largest single concentration of pastoral vale land lay in the West Midlands. Some farmers lived in villages, but much more settlement was located in hamlets and single farmsteads than in the arable vales [57]. Moreover, much land had long since been enclosed, and further enclosure proceeded here without serious controversy. Agricultural improvement could, of course, always result in more arable being ploughed up, thereby reducing the inhabitants' dependence on grain brought in from elsewhere. But in general these pastoral vales continued in pasture (though they might diversify somewhat by, for example, adding orchards, or growing tobacco) because the demand for dairy produce, livestock and horses expanded steadily in the seventeenth century. Where they differed from the arable vales was in their early adoption of industries as by-employments, a development which had begun in the Middle Ages and gathered pace in this period [60].

The economic situation of all the vales, whether arable or

pastoral, was as varied as their farming regimes, and each played a distinctive part in shaping local farming choices. Not all clay vales had good access to distant markets, so for this reason alone one specialisation was often more compelling than another. Leicestershire, for example, had no suitable river to carry grain to markets and therefore its crops were fed to animals who then walked there [40; 41; 58]. The West Midland claylands lay close to a developing industrial region around Birmingham; hence dairying was a good strategy in the seventeenth century [34]. The eastern end of the Vale of the White Horse in Berkshire, as we have seen, had such good means of communication with London along the Thames that it continued to concentrate on grain rather than livestock, but at the western end of the same vale farmers preferred pastoral dairying.

Class divisions in clayland parishes also constrained choices. More than one manorial estate in a parish meant less effective lordship than in a region like the downlands, where one parish often had only one manor and a resident lord. Sometimes in a clayland parish of divided lordship all manorial rights had been sold, as in Wigston Magna in Leicestershire, and freeholders took command in the place [42; 27, *134ff.*, *141ff.*]. They generally preferred to encourage the settlement of more people, for newcomers meant a demand for housing which they were ready to supply. Ownership of land thus continued to be dispersed among many, including small men, although nothing could prevent purposeful men appearing on the scene and engrossing land as the chance arose.

The resulting social pyramid in clayland parishes might therefore be broadly or narrowly based [56], tall or flattened. Declining grain prices between 1650 and 1750 did not favour small farmers in common-field parishes which preserved arable systems of the traditional kind; such men were better placed for survival in vales that were more pastoral and especially those that were expanding their dairying activities; dairying was, indeed, a most suitable small farmer's speciality. But dairying also succeeded best in enclosed country, so again the pastoral vales had the advantage.

The arable vales witnessed most clearly the decline of the small peasant between 1650 and 1750 [56, *Ch. 4*]. But the gentry did not consider vale countryside quite so attractive for their residences as the downs and wolds, and so where peasant farming did survive strongly through the sixteenth century, it might persist until 1750, particularly if livestock farming was a strong component. An alternative strategy for survival was for farmers to take up industrial by-employments. These had proved highly successful in pastoral areas where labour was already underemployed. Now they expanded into parts of the arable vale country where arable farming was weakening its hold. Textile spinning, wool, linen and canvas weaving, stocking knitting and lacemaking were some of the many possibilities [42]. Such industrial enterprises, which expanded in number and geographical extent during the seventeenth century and conferred a modest prosperity on the pastoral vales, now rescued some of the arable vales, whereas small farmers in the more staunchly arable-oriented vales languished.

MARSHLANDS

The main marshland regions lay along the flat coastlands of eastern and south-eastern England, namely in Holderness (Yorks.) [22, 35-6; 52], along the Lincolnshire coast [22, 35-6; 59, 49-78, 142-58], and the Essex [15, 136-8; 33] and Kent coasts [22, 59-64; 15, 134-6; 28]. Clay soils, of rich loam or of a stiffer, heavier consistency, predominated and were reminiscent of the vales, but this country differed in having extensive grazing grounds. At some points along the coast, moreover, these were greatly extended in the sixteenth century when changing sea currents caused rapid silting and allowed hundreds of acres of new land to be imbanked.

Proportions of arable to permanent pasture varied according to situation, inland parishes having more ploughland, those on the coast more grass. But in all but north-east Kent the arable was liable to lie under grass for terms of years. Some parishes, and hence some farms, stretched long distances over both kinds of land or had separate grazings on detached pieces of marsh at a distance. This confused distribution of land types reflected a time in the past when the population had been smaller and people had roamed more freely from inland to coast.

The marshland economy had a pastoral side which presented similarities with that of the fens, but there were also differences, and its social structure was markedly at variance. Cattle- and sheep-fattening or dairying were major specialities: the most intensive fattening was found on the south Kent marshes; dairying was carried on in Essex, though the milking of ewes for cheese died out after 1660. But substantial grain-growing was also part and parcel of the farm enterprise. Usually it was carried out in two common fields, and produced not only wheat, barley and oats but also valuable crops of beans and peas that were needed to supplement grass as animal feed, and in the seventeenth century rapeseed. From Holderness (Yorks.) and Lincolnshire, wheat was supplied to other east coast regions; from the marshes of Essex, wheat went to feed London citizens and oats to feed their horses.

In north-east Kent [28], including the isles of Thanet and Sheppey, the arable was more intensively cultivated than in any other marshland, not only for wheat, barley and pulses, but also for vegetables and other special crops like hops and canary seed. Farmers could conveniently send their wares along the Thames to the London market. Here, indeed, some of the farming came close to market gardening, judged by the amount of hand labour that went in to dibbling seed and frequent hoeing.

The least amount of land given to the plough was probably found in the south Kent marshes though even there the harvests of grain and pulses were not insignificant [15, *134–6*]. But farmers from the Kent downlands made increasingly greedy claims upon marshland grazings, renting pastures for fattening their beasts and sheep in summer before sending them to Smithfield in the autumn. To this end many animals were bought in from Wales and Scotland, and herdsmen were settled in makeshift cabins on the marsh to watch over them. The parish authorities had good cause to complain of the non-residents occupying their land and contributing nothing to the life of the community.

The same intrusion by non-resident graziers was a characteristic of the Lincolnshire marshlands [59, *148–51*]. It was most noticeable in the seventeenth century when livestock farming offered greater rewards, and wold farmers sought more summer grass to accommodate their larger interest in livestock. But when once they learned to grow turnips and clover in rotations on their own arable fields, they ceased to want the permanent pastures of the marshland and in the early eighteenth century withdrew.

In all marshland country society conformed to the conventional stereotype of an hierarchical structure. All classes were represented in their expected proportions. Gentlemen lived there as well as substantial yeomen, husbandmen and labourers. In other words, the social classes spanned the same range as in the clay vales. How far the small farmers survived the economic and social pressures of the seventeenth century depended often on the accidents of local circumstances, such as those discussed in Chapter 3. In the Lincolnshire marshlands as a whole, the signs point to a decline of the small proprietor as native yeomen and outsiders forged ahead and consolidated their land holdings. But the evidence is more impressionistic in other marshlands; in north-east Kent hard-working family farmers seem to have prospered; generalisations cannot yet safely be made.

HEATHLANDS

The hallmark of heathland is its sandy, gravelly soil, where grass grows thinly, heath and furze grow luxuriantly, and sheep are indispensable not only for fertilising but for treading and thus consolidating the arable soils. The most extensive areas of heathland lie in East Anglia [15, 72–80; 23, i, 198, 207–26], but a narrow strip of heath occupies the Bagshot, Ascot area of Surrey and the same soils underlie the New Forest in Hampshire [23, i, 343; 15, 82–3; 62], Sherwood Forest in Nottinghamshire [15, 72–83, 87–9, 165–9], and Charnwood Forest in Leicestershire. For reasons of agricultural practice, these forests are included here as heathlands. But other forests on clay land are separately classified below (p. 49), and the account given there of their social structure applies to these sandy forests equally with those on clay soils. Smaller areas of heathland are found, serving as commons in their respective parishes, as for example, on the south London plateau, at Blackheath, Bexley, Chislehurst and so on [23, i, 288–9]. Indeed, in most heathland regions, farmers had varying amounts of other kinds of soil, which helped them to survive on unpromising land, and which diversified their systems in detail [15, 77–80]. Near the Norfolk and Suffolk coasts, for example, they had marshes, and on the western side of Norfolk fenlands, both of which enabled them to fatten cattle or keep dairies. In the Bagshot area and in the New Forest they had woodland and could fatten pigs on the beechmast. In the New Forest they kept many ponies. In Sherwood

Forest they relied on industrial by-employments. All devoted the worst of their land to rabbit warrens, a form of profitable land use not to be despised.

All these activities, however, were carried on alongside, and integrated with, the same arable system; the arable land had to be fertilised by sheep that grazed on the grasslands by day and were folded on the fields at night. The grazing was thin and poor, but for that very reason it gave the mutton a sweet taste and produced a fine wool. The stronger arable soils could tolerate an ordinary three-course rotation, but these cases were rare. In Norfolk a two-course was possible, of grain and a fallow, followed by peas or oats and a fallow. But most usual was a system of convertible husbandry, by which farmers took two, three, or four crops only, followed by some years of grass leys. The main crop was barley, but occasionally rye, which likes a gravelly soil, was given pride of place. Buckwheat was not a common crop anywhere in England, but it was to be found on these heathlands, and it was specially valued in Norfolk for fattening poultry.

The heathland that was normally grazed was also treated as an outfield, from which segments were taken in for a few short years of cropping, five or six years in Sherwood, no more than one year in parts of Norfolk. Oats or skegs (a form of oats) were a common crop in these circumstances. Except in districts where marsh or fen was at the disposal of heathland farmers, their meadow was of poor quality. For fodder they grew vetches, lentils, tares and peas in the arable rotation and later, as the new legumes established themselves successfully, clover, trefoil, ryegrass, sainfoin and lucerne.

To make the most of hungry soils required much ingenuity. Manorial lords on the brecklands had long before established fold-course rights that were peculiar to East Anglia [26]. These required tenants to put their sheep on the lord's arable to fertilise it, and give him prior claim to the grazing on the fallows. Elsewhere ordinary common-field regulations were in force, and persisted to the end of the period. But enclosure went ahead steadily everywhere, accompanied by other improvements achieved by marling and fertilising with whatever manures were available. These measures enabled farmers to diversify somewhat in the seventeenth century, and grow special crops like liquorice in the Worksop district of Nottinghamshire, weld (a yellow dye plant) on the

Kentish downlands, fruit, hops, carrots on the Suffolk sandlings or – a choice for gentlemen landowners – timber trees and coppice. Still more significant for the future was the freedom given by enclosure to farmers on the better loams of east Norfolk to devise their famous Norfolk rotation, which made use of turnips and clover in order to strike a balance between the production of grain and livestock fodder and maintain the soil's precarious fertility. This rotation had variations, but it was based on the four crops: wheat, turnips, barley and clover. The clover was left down for one to four years, and sometimes a crop of barley was taken between the wheat and turnips. But the intention was to take one cleaning, or soil-improving, crop for every exhausting one. This rotation probably evolved out of the traditional system of taking crops for two, three or more years and then resting the land under temporary grass. Meanwhile, on the heathland that was treated as outfield, close-folding and marling gradually enabled more and more land to be brought into cultivation for longer periods – of four, six or even seven years of crops – thereby preparing them also for eventual enclosure and the regular Norfolk rotation.

The agricultural economy of the heathlands was in most places intricately intertwined with industrial by-employments such as gravel-digging, textile-weaving, stocking-knitting, and wood-working [23, 1, *307, 308*]. These helped small men to eke out a living, but such possibilities also attracted landless poor to the heathlands at a time when the commons in other areas were diminishing by enclosure and improvement. As the home of squatters, therefore, some of the heathlands, like the forests, tended to become synonymous with poverty. But this was not true everywhere, and some refined distinctions may emerge from more detailed research into heaths located in different areas and with different social structures. For example, restrictions on the ploughing of land in royal forests had inhibited settlement in earlier centuries, and in the period 1500 to 1750 one of these, the New Forest, seems to have been a stronghold of smallholders rather than paupers [23, 1, *344*]. In other areas, like Sherwood Forest, great landowners had large estates and exercised control to preserve a hierarchical society, not permitting the influx of an undue number of poor commoners. On the corn-sheep lands of Norfolk, villages were the characteristic form of settlement, over which gentlemen presided, with a keen eye to maintain their foldrights and a tenurial structure of farms that did not admit an excess of labourers. Squatters' settlements did appear, of course, and defied removal. But it is noticeable how many more such complaints came from the forests that lay on heavy clay soils rather than those on sandy heathlands [22, *409ff.*, *445ff.*, *463ff.*].(10)

FORESTS

Most forests were situated either on heavy clay soils [22; 23, i] or on sandy, gravelly heaths. They were numerous and scattered. Because of their extensive areas of scrub and improving grass, they are often called 'woodpasture' areas. They gave refuge to a great many people whose living depended on their common rights rather than a secure tenancy of land [62]. Historians have shown a revived interest in these forest areas in recent years, exploring their distinctive social structure and the mental attitudes of their inhabitants, as much as their farming [22, Ch. vii, *396ff.* especially *463ff.*; 39; 47; 31]. This has strengthened their claim to be considered as a distinct agrarian category. But in *The Agricultural Revolution*, where the basis of classification is the farming system, Professor Kerridge does not distinguish the forests from the surrounding vale lands. Therefore many forests lie in the large region of his Midland Plain. Moreover, because his regions are constructed on an eighteenth-century base, emphasis is laid on their *meagre* rough grazings [15, *99*], whereas other historians, who take their stand in the sixteenth century, see such common grazings as the lifeline of these communities, even though they were shrinking towards the end of the period.

Forest dwellers tended to live in hamlets as often as in villages, and being out of sight of most travellers, and unrestrained by resident gentlemen, they enjoyed a freedom that attracted the landless and luckless poor. Such arable land as lay in the forests might lie in common fields, but much was likely to be already enclosed in small parcels, and more could be contrived by intakes from the forest which went unopposed. Ploughland afforded food for the household but not more than that. Forest economies on the clays were pastoral (for the sandy forests, see above, p. 46), and concerned themselves with livestock of all kinds. Pigs were reared and fattened, cattle and sheep were reared and sent elsewhere to be finished for the butcher, horses were bred for work in the pits or on the roads. Where grassland clearings had substantially reduced the woodland, dairying [53; 54], even fattening [39], was established. Extra foodstuffs were gathered from the hedges or secured by poaching deer, hares or rabbits. Timber for fuel was, of course, plentiful, by-employments were numerous: they included woodworking, mining, ironmaking, metal-working, clothmaking, lace-making, blanket-making, leather-working and stocking-knitting.

Forest economies were well designed to keep people alive but not to make them rich men. Their activities dovetailed neatly with those of neighbouring, usually arable, regions; they supplied, for example, young cattle and sheep to be fattened elsewhere. But their large areas of underused, often rough, pasture and neglected woodland could not escape the attention of the sixteenth-century improvers, wishing to put more land under the plough. Disafforestation of the royal forests and clearance of other private woodland therefore proceeded steadily during the seventeenth century, and was accelerated during the Commonwealth period.(11) Usually this resulted in the upgrading of only select portions of woodland, as for example in Ashdown Forest in Sussex, since many poor woodland soils resisted improvement, even though lime was energetically bestowed to counteract their acidity. But remarkable success was occasionally achieved, so that some land labelled 'forest' in the sixteenth century was more accurately described as 'pastoral vale land' by the mid-eighteenth century.(12) In most such cases the new speciality was dairying, which needed good-quality grassland and some expansion of the arable acreage as well. Such a transformation occurred in the Forest of Arden in Warwickshire, where dairying served the adjoining industrial country around Birmingham and enough land was put under the plough by 1750 to end the forest dwellers' dependence on grain supplies brought in from outside [53; 54]. Commercial success of a new order in these forests fostered commercial attitudes, and some pasture farmers built up considerable wealth. At the very least, favourable price trends in the seventeenth century for livestock products preserved many forest economies. Nevertheless, incomers constantly aggravated the problems of poverty. Population pressure thus resulted in more woodland clearances and more recourse to by-employments. The forests became a byword for lawlessness and roguery. The inhabitants, however, valued the variety of resources and their freedom to use their ingenuity. The poor better equipped for survival than in arable country.

FELLS AND MOORLAND

Fells and moorland cover a large tract of land in northern England, extending through the Pennines into the border country, and east to include the north Yorkshire moors. Another region in the south-west encompasses Dartmoor, Exmoor and Bodmin moor and all the foothills, so that 'moorland farming' is the term usually used to describe practices in a large continuous area of north Cornwall and Devon [15; *149–54*; 22; *71–5, 76–8*; 23, i, *360ff., 375ff.*]. More detailed studies in the future will almost certainly refine our perceptions of these extensive, and seemingly uniform, regions, which have not yet received the same attention as the lowlands.

The acid soils of the moorland and a wet cold climate meant poor grassland and meagre chances to grow crops. Hardy livestock were the commercial mainstay of farmers, and the best land in the valleys was used for meadow to provide precious hay for their winter keep. An unkindly climate for grain made oats the first cereal for bread and beer, bigg – a four-rowed variety of coarse barley – the second, and rye the third. The grain harvest was no more than adequate for household needs.

The arable land could not bear crops for long sequences of years, and so one satisfactory system was 'infield-outfield', under which the infield was manured intensively while in use, and the outfield ploughed up for only a short period of years before returning to grass. In some places, however, small common fields had been laid out where a two- or even a three-course, rotation was followed. But since many people lived in hamlets or isolated farmsteads more often than in villages, and the acreage of ploughland was small compared with the vast expanses of grass, common-field management was inappropriate.

Moorland farmers, then, were first and foremost pastoralists, keeping cattle on the lower hillsides and sheep on the fells above. Very rarely they also kept goats. They had very generous common rights and made the most of the spring and summer grass to breed and rear. This meant that in some places on the Scottish border animals were taken up to sheilings in the higher mountains for the summer months. Elsewhere the moorlanders allowed lowlanders to bring in their cattle and sheep to 'agist', that is to feed, on grasslands that were surplus to their own needs. Most of the cattle and some of the sheep were then sold to lowlanders in the autumn for winter keeping.

Specialisation in livestock worked well, for the system was integrated with the regimes of vale farmers who bought the store animals. Both cattle and sheep were hardy, and having survived the fell conditions they throve in the gentler environment to which they were transferred. Moorland mutton that did reach the butcher direct from the fells was said to have a distinctive and fine flavour, particularly if it had been fed on young holly shoots. Wool from moorland sheep was readily bought for the textile industry. In the more sheltered dales and valleys dairying was possible, as in Wensleydale and Swaledale. Footsure ponies were, of course, essential for carrying loads from the fields to the farms and from farms to markets.

The population was thinly settled in the moorlands and people were accustomed to a hard life, but they enjoyed great personal freedom, and held their land by secure tenures that were as good as freeholds. Moreover, they had a variety of by-employments to eke out a living: mining or quarrying were one possibility, since many minerals lie underground in this kind of countryside; cottage handicrafts were another. These activities could readily be fitted into the farm routine, for it was not nearly so labour-intensive as in arable countryside [24].

The combination of farming with industrial employment enabled families to provide a living for more than one son, and so in some parts of the moorlands the custom still continued through the sixteenth and seventeenth centuries of dividing holdings to further this end, though it was never followed to impractical lengths [24]. But subdivision of land inevitably increased dependence on industrial occupations, and as industrial enterprises such as coal and ironmining expanded in the seventeenth century, some moorlanders in the north were more accurately described by 1750 as industrial workers with a subsidiary interest in farming than the reverse.

Life was precarious in the moorlands, as deaths from starvation during the severe years of food scarcity, 1587–8, 1597–8 and 1623, bear eloquent witness [1]. But people were deeply attached to their native places. Loyalties to family, neighbours and lord were the more intense because of the moorlanders' isolation. And they had much to be thankful for. In certain outward appearances their lives resembled those of yeomen in other parts of England. They lived in solid stone houses and commanded extensive grazing lands. Their standard of comfort was not nearly as high as the lowlanders' but competition for land was not severe, and the demand for their livestock expanded greatly in the seventeenth century. Circumstances therefore offered increasing scope to ambitious men to become great graziers with 250 head of cattle and 1000 sheep to their credit.

Seventeenth-century developments emphasised the advantages of the pastoral system in more than this one respect. Small farmers survived well; farms remained small – 10 acres was still a fair-sized holding, of which between 4 and 8 acres grew oats and bigg. But persistent efforts were made to extend the arable acreage somewhat, so that after 1623 bad harvests never again produced the high mortality seen earlier in Cumbria. Such improvement depended on enclosure which usually proceeded with agreement or, at least, without serious demur. But many enclosures were used to improve grassland rather than create more arable. The principal objectives were still cattle- and sheep-rearing and dairying. But the scale of enterprise and the details of stock management changed somewhat. Liming was more commonly practised, whereby the improved grass made it possible to winter more livestock [71]. (13) Some graziers even bought Scottish cattle in order to make the most of this opportunity. At the same time the droving of cattle from Scotland and Wales to London placed some moorland districts on a well-trodden route which established their own cattle dealings on a sounder footing in the national scheme of things. They were made even more secure when the government in 1663 and 1667 passed the acts prohibiting the import of young Irish cattle, thus guaranteeing to English breeders a higher price for animals that now became a much more scarce commodity. One of the moorlanders' misfortunes, however, was steadily falling wool prices. Men had less incentive to keep sheep and so these now took second place in their reckoning. But, all in all, the fells and moorlands were able to turn seventeenth-century changes in agricultural prices and the expansion of industrial enterprise to good account.

The fenlands were extensive around the Wash [15, *138ff.*; 22; *38ff.*; 23, I, *197ff.*; 49], around the estuary of the Humber [51], where several rivers including the Trent converge and a large arc of land is enveloped in the Isle of Axholme and Hatfield Chase [61; 23, I, *48*], and in the Somerset Levels in the neighbourhood of Sedgemoor and Bridgwater [65; 22, *78–9*; 23, I, *358ff*]. Rivers in all these areas overflowed their banks regularly in winter, and hundreds of acres of land were submerged. But the silt thereby deposited enriched the soil which produced lush grass in summer. This, therefore, was pastoral country, criss-crossed by dykes that served as lines of communication rather than barriers. Livestock were the fenlanders' main concern. They fattened cattle or supported dairies, kept sheep, if their land included some saltmarsh which prevented footrot, and bred horses.

The arable land was lifted out of range of the floods, and was very fertile. Continuous cropping without fallows was possible and so a small allotment of arable – no more than 10 acres, and sometimes half that – was enough to support a peasant family. It often lay in strips, but did not necessarily submit to common-field regulations. Relying mostly on their abundant grazings and unstinted common rights, which were the envy of more crowded vale parishes, the fenlanders made good use of other resources such as fish, wild fowl, osiers and willow for baskets, reeds for thatch, and reeds and peat for fuel. As in other pastoral regions, the inhabitants earned extra cash from by-employments, notably hemp and flax-weaving, since both crops, especially hemp, grew well here.

The fenland economy had adapted itself sensibly to the environment and, while feeding its own inhabitants satisfactorily, it also supplied some necessary foodstuffs to other parts of the kingdom, especially London. But in the sixteenth century when efforts to expand arable farming were at their height, such a seemingly benighted countryside could not be left in peace. Schemes were laid to drain and then plough the fens, which were implemented in the seventeenth century when economic conditions no longer favoured such a transformation. Pastoral economies were by then usually more prosperous than arable ones. The drainage caused much discontent and hardship to the inhabitants, who seized the chance during the civil wars to turn back the clock [45]. When work resumed after 1650 the newly drained fen tended to be used experimentally, for newer crops like rapeseed, dyeplants, teasels (in Somerset), and vegetables, while the lands that continued in the hands of the native fenlanders supported cattle, horses, sheep and geese as before.

The economy of the fens suited the people who lived there, even if it did not please outsiders. Its inhabitants were small or middling peasants without pretensions to wealth, inoculated by their upbringing against the ague (i.e. malaria) which dealt death to so many newcomers. They lived in villages standing slightly above the wet fen or in hamlets, but in neither place were they much disturbed by gentlemen, who did not choose to live in such an uncongenial climate. But because of its diverse resources and plentiful tracts of neglected land, the fenland attracted many landless poor in search of refuge. It continued to do so to the end of the period considered here. But seen in longer perspective, it was an ancient pastoral region in process of being converted into an arable one. By 1750 this transformation was no more than half completed.

5 Agricultural Innovations: An Agricultural Revolution?

In identifying agricultural regions, and following the course of their development for two and a half centuries from 1500 to 1750, we have noted modifications in farming specialities and social structures. All kinds of influences were at work to cause these changes. Among the most immediate and direct were the opportunities offered by new crops and new farming techniques. Most are enumerated in Kerridge [15, *14*, *181–325*, as follows: (i) 'up and down husbandry', meaning a rotation of crops for between two and nine years, but mostly of three to four years, followed by grass leys lasting usually between six and twelve years; (ii) fen drainage, which enabled land formerly flooded by swollen rivers to be dried out and used for arable (Kerridge also includes under this heading the imbanking of coastal marshland to protect it from the sea, and again, ultimately, to create more arable land); (iii) the use of more and varied fertilisers, including seaweed, seasand, industrial and urban waste, marl and lime, and so on; (iv) the floating of water meadows, by which a stream was deliberately led through channels and sluices to overflow the meadow for a period of weeks or months, thus permitting the deposit of silt to enrich the grassland, protecting the grass from frost, and hastening the growth of grass that was especially valued in early spring; (v) new crops, including rapeseed for oil, fodder crops like clover, sainfoin, lucerne and turnips, many vegetables including potatoes, and dye-crops; (vi) new systems to accommodate the new crops in arable rotations and allow some of them to be fully exploited for feeding animals; and, finally, (vii) new stock, though Kerridge does not claim that the breeds changed much, apart from the Cotswold and Midland pasture sheep, but they were certainly improved by better feeding.

The evidence found for these innovations in the sixteenth and seventeenth centuries is sufficient to leave no room for argument.

It means that many of the technical advances that used to be associated with the agricultural revolution in the century between 1750 and 1850 were under way 200 years before. But how rapidly did they spread? Here a lively debate begins because historians differ in their final verdict concerning the pace of progress. They have placed further difficulties in the way of general agreement on the matter by introducing the notion of an 'agricultural revolution' in the sixteenth and seventeenth centuries. That description was attached a century ago to the period of 1750–1850, admittedly at a time when historians had not properly explored the scale of improvements in the previous two and a half centuries. Should that dramatic title now be applied to the earlier period? Professor Kerridge claims that it should. In his book [15, *15*] his first sentence runs thus: 'This book argues that the agricultural revolution took place in England in the sixteenth and seventeenth centuries and not in the eighteenth and nineteenth.'

Agricultural historians of the eighteenth and nineteenth centuries have responded to this bold challenge by acknowledging that most of the technical advances associated with their agricultural revolution did, in fact, begin in the sixteenth and seventeenth centuries [77, *123*]. But agricultural changes are slow to achieve their full potential. So Professor Mingay sums up the now admittedly longer history of agricultural improvement thus: 'Considered in the broadest sense, the development of modern farming can be seen as stretching back into the sixteenth and seventeenth centuries, gathering pace in the later eighteenth and nineteenth centuries, and proceeding at its fastest in the present century' [76, *1–3*]. In other words, far-reaching agricultural change is here set within a much longer timespan, though the use of the term 'agricultural revolution' is still deemed appropriate for the period 1750–1850 because of the 'gathering pace' of progress in those years. Kerridge's reply to this argument is to dub it 'gradualist', spreading out the revolution 'over an impossibly long stretch of time' [72, *469*]. But the historian's stock in trade is change over long stretches of time; and indeed, another agricultural revolution, or the same one prolonged, has been suggested for the period 1815–80 [89].

A ready end could be put to an otherwise unending controversy if the concept of an 'agricultural revolution' were totally abandoned, and instead improvements were analysed as a continuum, to be divided between periods of more and less rapid change. Changes in the market are constantly stimulating, prodding or buffeting farmers to modify their procedures. But the pressures are stronger at some periods than at others, as experiences between 1500 and 1750 themselves bear witness. The sharp rise of population in the sixteenth century greatly encouraged grain-growing. The slump in grain prices between 1650 and 1750 cooled interest in that objective, but slowly promoted interest in special crops [23, 11]. This enthusiasm then evaporated when population rose again after 1750, and farmers returned to grain and livestock once more. Special crops revived yet again in the great agricultural depression at the end of the nineteenth century, from 1879 onwards. In each phase many of the same crops and systems were reviewed, but each time some progress was made in the details of management or their geographical spread.

Viewed at closer quarters, every new way in farming can be seen proceeding in well-defined stages. It never attracted supporters in an instant. A new crop or a new method had first to be tested by courageous pioneers who were prepared to lose money if they failed. It had to find the right soil and climatic conditions to succeed and, in the sixteenth century, these were far from being self-evident. Attempts were made to grow rapeseed in Kingswood Forest, near Bristol, for example, when in the end the best places proved to be on the drier eastern side of England. The financial and labour requirements of innovations had to be ascertained, for these suited one class of farmer rather than another. A gentleman might have the cash to prepare land, construct new buildings and employ a large labour force, which a yeoman farmer could not contemplate. Contrariwise, a very labour-intensive crop that required little land, like tobacco, might be taken up by small family farmers because both land and labour needs fitted their circumstances admirably, while the heavy labour needs deterred the husbandman, yeoman and gentleman [23, 11, *Ch. 19*; 86].

In short, new agricultural crops and methods could not find their niche within varied farming systems without undergoing a long process of trial and error that could involve many delays and setbacks. Using the term 'agricultural revolution' to describe the process arouses expectations which agricultural procedures could never meet. Changes were always put into effect slowly, moving hesitantly in widely separated places, and sometimes retreating again, temporarily or permanently. The historian only perceives a fraction of the false beginnings and frustrated hopes before an innovation establishes itself successfully.

Mark Overton has criticised Kerridge for exaggerating the spread of innovations. 'It seems that a few precocious farmers growing new crops, or dabbling in some new husbandry practice, provide sufficient grounds for claiming a revolution' [81, 122]. But this too is an exaggeration, for Kerridge spreads the revolution over two and a half centuries. Nevertheless, two and a half centuries were still not time enough for all the innovations to achieve their full potential. In Kerridge's view all delays and setbacks had been ended and overcome by 1750, and innovations had achieved all the progress possible by 1750. 'The reader may well wonder,' he writes at the end of his book, 'why more agricultural advances were not made in the last quarter of the eighteenth century and the first half of the nineteenth. The answer is no major innovation was possible, and little room remained for further improvement, for the simple reason that all the opportunities for such had previously been exhausted' [15, 338]. And again on the subject of turnips: 'If the cultivation of turnips made little progress after 1750, this was not, as has been assumed by some, because it was only just starting, but because it had already gone almost as far as it possibly could' [15, 339]. But turnips, in fact, are a good example of an innovation that passed through many phases of development, and did not make their full impact until long after 1750. In 1801 no more than 8.5 per cent of the recorded cropped area of England and Wales (excluding land under temporary grasses) grew turnips. But better varieties and improved drainage and field equipment permitted more varied management systems, and enabled them to spread into many new regions thereafter, until, by 1870, they occupied 18.2 per cent of the cropped acreage [79].

The debate on which centuries can claim to have seen an agricultural revolution has been lively and stimulating, but it has not shifted majority opinion towards the notion of an 'agricultural revolution' in the early modern period [90, 327]. Some have even regretted the use of the term 'revolution' at any period; Mark Overton dubs it 'thoroughly confusing' [81, 123]. Nevertheless, it is agreed that significant changes were in progress in the sixteenth and seventeenth centuries. If, then, we set aside the concept of revolutions, and also bear prominently in mind the diversity of opportunity and practice in England's many farming regions, we may describe agricultural changes in the period 1500–1750 in more cautious, if less colourful, terms and still concede heroic achievements. Agricultural production and, by implication, agricultural efficiency improved remarkably between 1500 and 1750. Farmers' success in feeding the population and avoiding serious shortages after 1623 was achieved in three ways: (i) by the more careful cultivation of arable and selection of stock, following traditional lines; (ii) by an extension of the acreage under cultivation through reclamation and the gradual upgrading of land, much assisted by enclosure, which gave farmers more freedom of manoeuvre; and (iii) by more productive and varied farming systems, exploiting the innovations listed in Kerridge's study [15, *14*]. But among these three ingredients, our documents do not allow anything but guesses concerning the proportionate contribution of each in the final result.

More careful cultivation along traditional lines – factor (i) – was made possible when gentlemen took an active interest in farming, when more books of practical advice appeared, were read and acted upon after the mid-sixteenth century, and a still wider exchange of practical experience occurred in the course of the seventeenth century [87, *295–301*]. Eventually, in the eighteenth century, books were supplemented by journals, and farmers' clubs were established. The second mode of improvement – factor (ii) – was achieved through increases in the acreage under cultivation, which enclosure actively promoted. Drs Bowden and Thirsk rate these as of more significance than the other two [22, *199, 606–7*], but the resulting increase in food production cannot be measured. It may be worth noting, however, that even in the period 1700–1850, two-thirds of the increase in agricultural output is tentatively attributed to this same factor, the increase in cultivated acreage being mainly a consequence of enclosure [76, *10*]. The third factor was the new crops and new techniques, already discussed. Their importance continued to be underlined by eighteenth- and nineteenth-century writers, observing developments in their own day.

When advance in the eighteenth and nineteenth centuries has to be accommodated alongside the evident progress of agriculture in the two centuries before, it is timely to remind ourselves again of the regional contrasts. In the four northern counties, for example, improvements were hardly beginning to be adopted before 1700, and most progress was made after 1750 [29]. In East Anglia, on the other hand, the spread of some improvements was evident from the 1580s [83, *209*], and it gathered speed noticeably after 1690. Yet it still left much room for improvement in the eighteenth century [74, *129*].

The historian has to accept the impossibility of measuring agricultural change in the past in a totally satisfactory way. The documents can, of course, always be differently, or more comprehensively, analysed [82; 83; 85] but they never offer a complete picture, and the final verdict on the scale of change has to be a matter of personal judgement. The current dictionary definition of a revolution does not insist on rapid change, but it does insist on radical change. To justify the use of this term in a twentieth-century context economists look for evidence of dramatic increases in productivity, either of land or of labour, since present-day statistics yield such information. But the past does not. And our knowledge of regional variety urges extreme caution in using local examples to make large general claims. It is all too easy to treat southern or eastern England as though it were the whole kingdom. As we have seen, every region proceeded at its own pace of change. That pace was determined by the natural agricultural attributes of each area, and the current possibilities of improvement. It depended on the existing allocation of land between the classes, and the size and structure of the labour force. Most of all, it responded to the economic opportunities afforded by the markets, and by the transport facilities then available to reach those markets. The sum total of the varied regional improvements carried through after 1500 enabled farmers to feed about 3 million more people in 1700 than in 1540, and almost 20 million more in 1880 than in 1750 [83, *328*]. But the second phase of effort gathered momentum during the first phase. In different contexts, and starting from a different base, both achievements represented heroic endeavour.

Glossary

To enclose – to fence, hedge or otherwise enclose a piece of ground which has hitherto lain open, either in the arable fields, the meadows, or on common pastures. The principal purpose of enclosure was to restrict, or more usually extinguish altogether, common rights of grazing enjoyed by others over the land. Other advantages of enclosure lay in the fact that the hedges, walls or fences offered shelter to livestock in bad weather, and trees in the hedges afforded additional timber. Often enclosure was followed by the conversion of arable land to pasture, and in the sixteenth and seventeenth centuries, and also since, the term 'enclosure' has been loosely used to encompass both processes. In fact, it was perfectly possible to enclose without permanently changing the use of the land. Enclosure still had advantages because it gave the farmer more freedom in the choice of crops. He did not have to allow the commoners' animals to graze his land by certain dates, after harvest and in fallow seasons.

Foldcourses – these are peculiar to Norfolk, north-west Suffolk and Cambridgeshire. Instead of common rights belonging to landholding members of a whole community, they belonged to the manorial lord or his lessee. But as East Anglian townships usually had more than one manor, more than one demesne flock had to be accommodated on the available grazing. Hence each flock had its own foldcourse, a defined area which provided grazing at the various seasons. In detail, practices varied: in some townships some common grazing was reserved for the cattle of the peasant farmers, in others the grazing was wholly divided between the foldcourses, that is, it belonged to the lord. For further explanations and examples, see [2, *314–22*].

Marl – was a substance dug from various depths beneath the land surface, thus creating marlpits that can still be identified in the fields. Marl was either more calcareous or more argillaceous (clayey) in varying proportions, and was used appropriately to lighten clay or strengthen sandy soils, thus improving their fertility.

Notes

1. Donald Gibson (ed.), *A Parson in the Vale of White Horse. George Woodward's Letters from East Hendred, 1753–1761* (Gloucester, 1982), p. 47.

2. A.D. Powell, *John Aubrey and his Friends* (1948), pp. 38–9. 'Milk meats', or more usually 'white meats' was the term used for dairy produce.

3. I regret having used the term 'mixed farming' in vol. IV of *The Agrarian History* [22, 4] to describe arable-based farming in the sixteenth century. The term was chosen to underline the fact that such farming relied on animals to fertilise the fields, a practice that is not always deemed necessary in the twentieth century. But a clearer distinction is made by contrasting 'arable' with 'pastoral' systems, and those are the terms used here.

4. I wish to thank Dr Janie Cottis for allowing me to use her unpublished work in this section. It is contained in her Ph.D. thesis, Reading University (1984), entitled 'Agrarian Change in the Vale of White Horse, 1660–1760'.

5. I wish to thank Dr Malcolm Watkinson for allowing me to use as illustration of the argument in this section his unpublished M.Phil. thesis, Leicester University (1985), entitled 'Population Change and Agrarian Development: the Parishes of Bradley, Scartho and Humberston, c.1520–c.1730'.

6. The terms 'close' or 'closed' to describe parishes dominated by one landowner, in contrast with 'open' parishes, where ownership was divided between two or several owners, should not be confused with the terms 'enclosed' and 'open' referring to land in closes or in open (common) fields. The significance of 'close' and 'open' parishes is described in [13]. For practical guidance on methods of identifying numbers of landowners and then classifying individual parishes as 'open' or 'close', see D. Mills, 'English villages in the eighteenth and nineteenth centuries: a sociological approach. Part I, The concept of a sociological classification',

Amateur Historian, 6, no. 8 (1965), and Part ɪɪ, 'A survey of the main types of source material', *ibid.*, 7, no. 1 (1966). For an example of such a classification of parishes in the division of Kesteven, Lincolnshire, see D.R. Mills, 'The Poor Laws and the distribution of population, *c.*1600–1860, with special reference to Lincolnshire', *Trans. Institute of British Geographers*, 26 (1959).

7. For a remarkable account by a seventeenth-century farmer of part of his working routine on this type of land on the Yorkshire wolds, see Donald Woodward (ed.), *The Farming and Memorandum Book of Henry Best of Elmswell, 1642*, British Academy, Records of Social and Economic History, New Ser., ᴠɪɪɪ (1984). For two brief extracts from this account, see [10; *123–6*].

8. For a description of Lincolnshire wold farmers in 1623, temporarily in acute difficulties when grain and wool prices fell together, see [10, *24*].

9. For a seventeenth-century description of this system, see [10, *134*]. For a lively debate in 1656 for and against enclosure in this kind of country, see [10, *144–50*].

10. Professor Everitt in this chapter on labourers treats all the forests together, without distinguishing the labourers' conditions in each type.

11. For arguments in 1612 for improving the forests, see [10, *116–20*], and again in 1653 [10, *135–40*].

12. Hence 'woodpasture' is omitted as a distinct category on the map in [23, ɪ, *xx–xxi*], and amalgamated with other pastoral types.

13. For liming and a general description of the farming regime in Wharfedale, West Yorkshire, in 1664, see [10, *151–3*].

Select Bibliography

Unless otherwise stated, the place of publication of books is London.

GENERAL BACKGROUND

[1] A.B. Appleby, *Famine in Tudor and Stuart England* (Liverpool, 1978). Illustrates food crises.
[2] A.R.H. Baker and R.A. Butlin, *Studies of Field Systems in the British Isles* (Cambridge, 1973). Describes regional variations in field systems.
[3] M. Beresford, *The Lost Villages of England* (1954. Repr. 1983). The varied regional distribution of deserted villages is significant in this general survey.
[4] C.G.A. Clay, *Economic Expansion and Social Change*: *England, 1500–1700. Vol.* I, *People, Land and Towns; vol.* II, *Industry, Trade and Government* (Cambridge, 1984). A wide-ranging survey by a specialist in agrarian history.
[5] L.A. Clarkson, *The Pre-Industrial Economy in England, 1500–1750* (1971). A well-informed, careful survey.
[6] D.C. Coleman, *The Economy of England, 1450–1750* (Oxford, 1977). An especially helpful survey in distinguishing trends before 1650 and after 1650.
[7] B.A. Holderness, *Pre-Industrial England. Economy and Society, 1500–1750* (1976). Another good survey by a specialist in agrarian history.
[8] Lord Ernle, *English Farming Past and Present* (new sixth edn., 1961).
[9] R.H. Tawney, *The Agrarian Problem in the Sixteenth Century* (1912).
[10] Joan Thirsk and J.P. Cooper (eds), *Seventeenth-Century Economic Documents* (Oxford, 1972).

[11] H.C. Darby, 'Some early ideas on the agricultural regions of England', *Agricultural History Review*, 11 (1954). An historical survey of attempts by writers from 1500 to 1860 to define agricultural regions.

[12] A.M. Everitt, 'Country, county and town: patterns of regional evolution in England', in Everitt, *Landscape and Community in England* (1985). Also in *Transactions of the Royal Historical Society*, 29 (1979).

[13] B.A. Holderness, '"Open" and "Close" parishes in England in the eighteenth and nineteenth centuries', *Agricultural History Review*, 20 (1972).

[14] W.G. Hoskins, 'Regional farming in England', *Agricultural History Review*, 11 (1954). A plea for more regional studies.

[15] E. Kerridge, *The Agricultural Revolution* (1967).

[16] J.D. Marshall, 'Why study regions?', *Journal of Regional and Local Studies*, 5 (1) (1985), 6(1) (1986). Thoughtful on the general issue with a commentary on some of the existing geographical literature.

[17] G.E. Mingay, Review of E. Kerridge, *The Agricultural Revolution*, *Agricultural History Review*, 17 (1969). Criticises the notion of another, earlier agricultural revolution.

[18] V. Morgan, 'The cartographic image of "The Country" in early modern England', *Transactions of the Royal Historical Society*, 5th Ser., 29 (1979). Describes the interest in England's regional diversity, and the use of county maps for decorative purposes in the sixteenth century.

[19] A. Rogers, 'Industrialisation and the Local Community', in *Region and Industrialisation. Studies on the Role of the Region in the Economic History of the last two Centuries*, ed. S. Pollard (Göttingen, 1980). A general discussion of industry in rural areas, based on a study of South Nottinghamshire, 1670–1720.

[20] L.D. Stamp, *The Land of Britain and How it is Used* (1946). A readable introduction by a geographer to England's regions and their changing land use.

[21] L.D. Stamp, *The Land of Britain – its Use and Misuse* (1950).

[22] Joan Thirsk (ed.), *The Agrarian History of England and Wales, Vol. IV, 1500–1640* (Cambridge, 1967).

[23] Joan Thirsk (ed.), *The Agrarian History of England and Wales, Vol.* v, *1640–1750. Part* I. *Regional Farming Systems; Part* II. Agrarian Change (Cambridge, 1984, 1985).

[24] Joan Thirsk, 'Industries in the countryside', in Thirsk, *The Rural Economy of England. Collected Essays* (1985).

[25] D. Underdown, *Revel, Riot, and Rebellion. Popular Politics and Culture in England, 1603–1660* (Oxford, 1985). Matches economic regions with their respective political loyalties and cultural traditions. See also [64].

LOCAL STUDIES

[26] K.J. Allison, 'The sheep-corn husbandry of Norfolk in the sixteenth and seventeenth centuries, *Agricultural History Review*, v (i) (1957).

[27] M.K. Ashby, *The Changing English Village. A History of Bledington, Gloucestershire in its Setting, 1066–1914* (Kineton, Warws., 1974). A wide-ranging study of a clayland vale community.

[28] D. Baker, *Agricultural Prices, Production, and Marketing, with special Reference to the Hop Industry, North-East Kent, 1680–1760* (New York, 1985). Describes an intensively cultivated region, exploiting new crops and its proximity to London.

[29] J.V. Beckett, 'The decline of the small landowner in eighteenth- and nineteenth-century England: some regional considerations', *Agricultural History Review*, 30 (ii) (1982). Discusses the general issue and the example of Cumbria.

[30] P. Brassley, *Agricultural Economy of Northumberland and Durham in the Period, 1640–1750* (New York, 1985). Describes and carefully documents regional differentiation within these two counties.

[31] C.W. Chalklin, 'The rural economy of a Kentish wealden parish, 1650–1750', *Agricultural History Review*, x (i) (1962). A good example of a woodpasture region.

[32] A.C. Chibnall, *Sherington. Fiefs and Fields of a Buckinghamshire Village* (Cambridge, 1965). A fine reconstruction of a vale community in north Buckinghamshire from Roman times to the nineteenth century.

[33] B.E. Cracknell, *Canvey Island: the History of a Marshland*

Community (Occasional Paper, no. 12, Dept. of English Local History, Leicester University, 1959).

[34] P. Edwards, 'The development of dairy farming on the north Shropshire plain in the seventeenth century', *Midland History*, IV, 3 & 4 (1978).

[35] Mary E. Finch, *The Wealth of Five Northamptonshire Families, 1540–1640* (Northants. Record Soc., XIX, 1956). Illustrates changes in farming specialities by gentlemen farmers in an East Midland county.

[36] A. Harris, *The Open Fields of East Yorkshire* (East Yorkshire Local History Ser., IX, 1959).

[37] A. Harris, The agriculture of the East Riding of Yorkshire before Parliamentary enclosures', *Yorkshire Archaeological Journal*, XL (1959–62).

[38] M.A. Havinden, 'Agricultural progress in open-field Oxfordshire', *Agricultural History Review*, IX (ii) (1961). A classic example of agricultural improvement in a Midland county.

[39] D.G. Hey, *An English Rural Community. Myddle under the Tudors and Stuarts* (Leicester, 1974). A fine study of a woodpasture parish in the West Midlands, populated by peasant farmers.

[40] W.G. Hoskins, 'The Leicestershire farmer in the sixteenth century', in Hoskins, *Essays in Leicestershire History* (Liverpool, 1950). Pioneered the use of probate inventories to identify a Midland county's farming specialities.

[41] W.G. Hoskins, 'The Leicestershire farmer in the seventeenth century', in Hoskins, *Provincial England. Essays in Social and Economic History* (1963).

[42] W.G. Hoskins, *The Midland Peasant. The Economic and Social History of a Leicestershire Village* (1957). A classic study of an East Midland parish, Wigston Magna.

[43] B. Jennings (ed.), *A History of Nidderdale* (Huddersfield, 1967). A good example of a fell parish with a rural industry.

[44] E. Kerridge, 'Agriculture, *c.*1500–*c.*1793', in *Victoria History of the County of Wiltshire*, IV (1959). Illustrates the varied fortunes of several farming regions in one county, emphasising especially the contrast of chalk and cheese countries.

[45] K. Lindley, *Fenland Riots and the English Revolution* (1982). Describes the disturbances accompanying the drainage, which began the change from a pastoral to an arable economy.

[46] G. Longman, *A Corner of England's Garden. An Agrarian History of South-West Hertfordshire, 1600–1850*, 2 vols (Bushey, Herts., 1977). A well-documented study of parishes on chalk and London clay, where clover and sainfoin were introduced after 1660.

[47] P.A.J. Pettit, *The Royal Forest of Northamptonshire* (Northants. Record Soc., xxiii, 1958). A valuable study of a forest economy in the East Midlands, which included rural industry.

[48] M.R. Postgate, 'The field systems of Breckland', *Agricultural History Review*, x (ii) (1962). Explains the practice of infield and outfield cultivation, with rotations of differing intensity, on these poor soils in East Anglia.

[49] J.R. Ravensdale, *Liable to Floods. Village Landscape on the Edge of the Fens, AD 450–1850* (Cambridge, 1974). A perceptive study of three fenland parishes, Waterbeach, Landbeach and Cottenham.

[50] A. Rogers, 'Rural industries and social structure. The framework knitting industry of South Nottinghamshire, 1670–1840', *Textile History*, 12 (1981). A careful study, parish by parish, of population and wealth in relation to the location of rural industry.

[51] June Sheppard, *The Draining of the Hull Valley* (York, 1958). Describes the drainage, but not its effect on agriculture or social structure.

[52] J.A. Sheppard, *The Draining of the Marshlands of South Holderness and the Vale of York* (East Yorkshire Local History Ser., xx, 1966).

[53] V. Skipp, *Crisis and Development. An Ecological Case Study of the Forest of Arden, 1570–1674* (Cambridge, 1978). A study of a forest economy undergoing considerable stress and change, presented in an original way as challenge and response.

[54] V. Skipp. 'Economic and social change in the Forest of Arden, 1530–1649', in Joan Thirsk (ed.), *Land, Church, and People. Essays presented to Professor H.P.R. Finberg* (*Agricultural History Review*, 18, Supplement, 1970). An analysis of a forest's agriculture and agrarian structure.

[55] M. Spufford, *A Cambridgeshire Community. Chippenham from Settlement to Enclosure* (Occasional Paper, No. xx, Dept. of English Local History, Leicester University, 1965).

Traces the decline of small peasant farms and amalgamations into larger units in a downland chalk parish.

[56] M. Spufford, *Contrasting Communities. English Villagers in the Sixteenth and Seventeenth Centuries* (Cambridge, 1974). A classic study of the contrasting histories of three parishes, Willingham in the fens, Chippenham on chalk downland and Orwell on clay.

[57] P. and M. Spufford, *Eccleshall. The Story of a Staffordshire Market Town and its dependant Villages* (Keele, Staffs., 1964). A short study of a pastoral parish containing many hamlets.

[58] Joan Thirsk, 'Agrarian history, 1540–1950', in W.G. Hoskins and R.A. McKinley (eds), *The Victoria History of the County of Leicester*, vol. II (1954). Dwells on general changes affecting the whole county rather than distinguishing regions.

[59] Joan Thirsk, *English Peasant Farming. The Agrarian History of Lincolnshire from Tudor to Recent Times* (1957. Repr. 1981). Divides Lincolnshire into four regional types, and follows changes in each, namely the fens, marshlands, clay vales, and chalk wolds and limestone cliff.

[60] Joan Thirsk, 'Horn and thorn in Staffordshire: the economy of a pastoral county', in Thirsk, *The Rural Economy of England* (1985).

[61] Joan Thirsk, 'The Isle of Axholme before Vermuyden', in Thirsk, *The Rural Economy of England* (London, 1985). Also in *Agricultural History Review*, I (1953). Describes a fenland economy.

[62] C.R. Tubbs, 'The development of the smallholding and cottage stock-keeping economy of the New Forest', *Agricultural History Review*, XIII (1965). Describes a forest economy.

[63] G. Tupling, *The Economic History of Rossendale* (Chetham Soc., LXXXVI, 1972). Describes a moorland economy with rural industry.

[64] D. Underdown, 'The Chalk and the Cheese: Contrasts among the English Clubmen', *Past and Present*, 85 (1979). Political activities in the Civil War are here related to agricultural and cultural regions.

[65] M. Williams, *The Draining of the Somerset Levels* (Cambridge, 1970). An account of drainage efforts which altered the old fenland economy.

[66] M. Zell, 'Population and family structure in the sixteenth-century Weald', *Archaeologia Cantiana*, C (1984). A good

example of a woodpasture community in southern England, in which population growth and mobility and the role of the textile industry are examined.

AGRICULTURAL INNOVATIONS

[67] J.H. Bettey, 'The development of water meadows in Dorset during the seventeenth century', *Agricultural History Review*, 25 (i) (1977).

[68] J. Broad, 'Alternate husbandry and permanent pasture in the Midlands, 1650–1800', *Agricultural History Review*, 28 (ii) (1980).

[69] H.C. Darby, *The Draining of the Fens* (Cambridge, 1940).

[70] F. Emery, 'The mechanics of innovation: clover cultivation in Wales before 1750', *Journal of Historical Geography*, 2 (1) (1976). An outstanding illustration of the way a new crop spread between identifiable people, connected by kinship or friendship.

[71] M. Havinden, 'Lime as a means of agricultural improvement: the Devon example', in C.W. Chalklin and M.A. Havinden (eds), *Rural Change and Urban Growth, 1500–1800. Essays in English Regional History in Honour of W.G. Hoskins* (1974).

[72] E. Kerridge, 'The agricultural revolution reconsidered', *Agricultural History*, 43 (1969). Explores the origins of the term 'agricultural revolution', used to describe the period 1750–1850.

[73] E. Kerridge, 'The sheepfold in Wiltshire and the floating of the watermeadows', *Economic History Review*, 2nd Ser., 6 (1953–4).

[74] E. Kerridge, 'Turnip husbandry in High Suffolk', *Economic History Review*, 2nd Ser., 8 (1955–6). Conclusion modified in [83].

[75] C. Lane, 'The development of pastures and meadows during the sixteenth and seventeenth centuries', *Agricultural History Review*, 28 (i) (1980). Documents a new interest in the quality of grassland.

[76] G.E. Mingay (ed.), *The Agricultural Revolution. Changes in Agriculture, 1650–1880* (1977). The introduction to this volume of documents discusses the validity of the term 'revolution'.

[77] G.E. Mingay, 'The "Agricultural Revolution" in English history: a reconsideration', *Agricultural History*, 37 (1963).

[78] G.E. Mingay, 'Dr Kerridge's "Agricultural Revolution": a comment', *Agricultural History*, 43 (1969). A robust contribution to the debate at a general level, in answer to Kerridge's article in the same journal.

[79] Raine Morgan, 'Root crops', in Chapter IV on 'Farming techniques', *The Agrarian History of England and Wales*, VI, *1750–1850* (Cambridge, forthcoming).

[80] R.B. Outhwaite, 'Progress and backwardness in English agriculture, 1500–1650', *Economic History Review*, 2nd Ser., XXXIX (1986). Stresses the downward pressures on average arable yields (e.g. through the cultivation of more marginal land), whereas most writers concentrate on the improvements.

[81] M. Overton, 'Agricultural revolution? Development of the agrarian economy in early modern England', in A.R.H. Baker and D. Gregory (eds), *Explorations in Historical Geography. Interpretative Essays* (Cambridge, 1984). A general discussion of change, but using Norfolk and Suffolk probate inventories to chart the introduction of root crops and grass substitutes, 1587–1729.

[82] M. Overton, 'Computer analysis of an inconsistent data source: the case of probate inventories', *Journal of Historical Geography*, 3 (4) (1977).

[83] M. Overton, 'The diffusion of agricultural innovations in early modern England: turnips and clover in Norfolk and Suffolk, 1580–1740', *Transactions of the Institute of British Geographers*, New Ser., 10 (1985). Uses large numbers of probate inventories to map the geographical spread of these two new crops and identify the social classes of innovators.

[84] M. Overton, 'English probate inventories and the measurement of agricultural change', in A. Van der Woude and A. Schuurman (eds), *Probate Inventories* (Wageningen, 1980).

[85] M. Overton, 'Estimating crop yields from probate inventories', *Journal of Economic History*, XXXIX (2) (1979).

[86] Joan Thirsk, 'New crops and their diffusion: tobacco-growing

in seventeenth-century England', in Thirsk, *The Rural Economy of England* (1985).

[87] Joan Thirsk, 'Plough and pen: agricultural writers in the seventeenth century', in T.H. Aston *et al.* (eds), *Social Relations and Ideas. Essays in Honour of R.H. Hilton* (Cambridge, 1983). Discusses, and gives examples of, the influence of agricultural writers on farming practice.

[88] Joan Thirsk, 'Seventeenth-century agriculture and social change', in Thirsk, *The Rural Economy of England* (1984). Also in Thirsk (ed.), *Land, Church, and People* (1970), see [54]. Contrasts different strategies, including the choice of different new crops, between regions in the seventeenth century.

[89] F.M.L. Thompson, 'The Second Agricultural Revolution, 1815–80', *Economic History Review*, 2nd Ser., xxi (1968).

[90] D. Woodward, 'Agricultural revolution in England, 1500–1900: a survey', *The Local Historian*, 9, no. 7 (1971). A very fair and balanced review of the literature on three agricultural revolutions, with good bibliography.

Index

Abingdon 33
Adderley 21
Agrarian regions 17, 25, 26,
  37ff.
Agriculture:
  change 57–61; stages
    of 58–60
  innovations 9, 56–60
  productivity 60
  regions defined 9, 11ff.,
    19–21, 23, 26, 37ff.
  revolution 9, 56–61
  specialisation, economic
    influences on 17–19;
    political influences 19;
    physical influences 12–15;
    social influences 15–17
Arable farming 64
  considered chronologically 10,
    16–17, 18;
    geographically 13, 14–15,
    27, 40, 41–2; socially 16
  in fens 29
  in Lincolnshire
    marshland 34–6
  in Midland Plain 25
  in Vale of White Horse 32–4
  on wolds and downs 40, 65
Arden Forest 50
Aubrey, John 21–2
Axholme, Isle of 27, 54

Berkshire 32
Birmingham 43, 50
Bradley, Lincs. 34–6
Buckwheat 38, 47
Butter 18

By-employments *see* Rural
  industries

Canary seed 45
Cattle Acts 19, 53
Cattle-fattening 16, 17, 19, 25,
  27, 29–30, 34, 50, 54
Cattle-rearing 14, 18, 27, 29–30,
  34, 49–50, 51–3
Cheese 18, 22
Cider 18
Civil War 18, 21, 29, 55
Climate 12
Closed parishes 35, 40, 64–5
Clover 34, 38, 41, 42, 47, 48, 56
Clubmen 21
Conversion, to pasture 10, 13,
  34, 41, 42
  to arable, 10, 13, 16–17, 18,
    54–5
Convertible husbandry 41, 47,
  56
Cornwall 51
Cottis, Dr Janie 32
Crop rotations 25, 26, 30, 34,
  40, 47, 51, 54
  Norfolk 48
Curdworth, John 21

Dairying 10, 14, 21
  and social structure 15–16,
    17, 43
  in fenland 27, 54
  in forests 50
  in Lincs. marshland 34–5
  in Midland Plain, 25
  in Weald of Kent 29–30